Principles of estimating

Trevor M. Holroyd

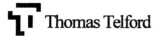
Thomas Telford

Published by Thomas Telford Publishing, Thomas Telford Ltd, 1 Heron Quay, London E14 4JD.

URL:http://www.t-telford.co.uk

Distributors for Thomas Telford books are
USA: ASCE Press, 1801 Alexander Bell Drive, Reston, VA 20191-4400
Japan: Maruzen Co. Ltd, Book Department, 3–10 Nihonbashi 2-chome, Chuo-ku, Tokyo 103
Australia: DA Books and Journals, 648 Whitehorse Road, Mitcham 3132, Victoria

First published 2000

Also available from Thomas Telford Books

CDM questions and answers: a practical approach. Pat Perry. ISBN 07277 2742 7
CDM regulations explained. Raymond Joyce. ISBN 0 7277 2034 1
CESMM3. Institute of Civil Engineers. ISBN 0 7277 1561 5
CESMM3 example. Martin Barnes. ISBN 0 7277 1657 3
CESMM3 handbook. Martin Barnes. ISBN 0 7277 1658 1
CESMM3 price database 1999/2000. Edited by E. C. Harris. ISBN 0 7277 2779 6
Site management for engineers. Trevor M. Holroyd. ISBN 0 27 2736 2

A catalogue record for this book is available from the British Library

ISBN: 0 7277 2763 X

Typeset by Gray Publishing, Tunbridge Wells, Kent
Printed and bound in Great Britain by MPG Books Ltd, Bodmin, Cornwall

Preface

The Documentation which I submitted for chartered status consideration included reports of knowledge and experience, design calculations and drawings, a bill of quantities and a priced estimate.

Whilst the preparation of the various parts took a good deal of time, the only major difficulty lay in the preparation of the bill and the estimate. Nothing that I had learned previously, either during my education or at work, had given me the slightest idea of what a bill or an estimate was.

As I gained experience it became quite clear that knowledge of quantities and pricing was of key importance to engineers who wished to become managers. For some 30 years I have encouraged every engineer in the teams which I have managed to become competent estimators. Lecturing to various organisations has enabled me to reach a wider audience.

Construction businesses survive by submitting accurate and well prepared tenders and by knowing what things cost. In writing this book I hope to bring some simple ideas to the notice of the reader which will, if correctly implemented, provide them with a key skill. A degree of practice will enable them to correctly apply the 'principles of estimating'.

Acknowledgements

Steve Arnold founded CTA Services in 1984 with the intention of providing relevant training by practitioners for practitioners. He invited people from within the Construction Industry to deliver suitable relevant training courses. My first proposal was for a course entitled 'Principles of Estimating'.

CTA Services delivered the courses. Now, as Thomas Telford Training, delivery of a widening course range continues under the auspices of the Institution of Civil Engineers. I am proud and privileged to contribute.

On behalf of myself, and those who have also benefited from this work, I would like to say 'thank you'.

I have written this book for Steve Arnold.

CITB
Bircham Newton
Kings Lynn
Norfolk
PE31 6LH

Tel: 01 485 577577
Fax: 01 485 577497

HSE Books
PO Box 1999
Sudbury
Suffolk
CO10 6FS

Tel: 01 787 881165
Fax: 01 787 313995

The Institution of Civil Engineers
1 Great George Street
Westminster
London
SW1P 3AA

Tel: 020 7222 7722
Fax: 020 7222 7500

Thomas Telford Ltd
1 Heron Quay
London
E14 4JD

Tel: 020 7987 6999
Fax: 020 7538 4101

Contents

1

Estimating from first principles

I have tutored many courses on the subject of Estimating and other similar topics and am amazed at how many of the delegates attending insist that they have no experience of the estimating process and that they do not know how prices are derived. Are they correct?

All of us, as individuals, families, social groups, businesses, regulate our lives and the actions we take by evaluating the cost of items or events and then deciding what we can afford to do. And what we can *not* afford to do.

If we wish to lead a pleasant and orderly life we work to a budget. We consider our income and plan our expenditure so that the two balance or, better still, we show a small surplus of income over expenditure. Income is the value people place on our efforts, expenditure is the cost of the operation.

Few of us would consider booking a holiday without first assessing the likely overall cost: our expenditure. Will our income allow this? That is, can we afford it? Will we be able to pay?

If we want to buy a house or to refurbish the home we live in, then estimates of cost will be the norm. We will take few gambles. Rather, we will take great care to estimate the price — how much it will cost — correctly.

Most families work carefully to keep their weekly household costs within acceptable limits. The successful unit keeps its costs within the limits it can afford — it will work to the budget plan. Unsuccessful units will have all sorts of problems. Their costs will be too high, expenditure greater than expected. They will have got it wrong.

I am sure you will agree that you *do* prepare estimates in your everyday life and that you realize how important it is to base the estimates on real costs and not guesswork.

Furthermore, if you are uncertain how much a holiday in a far-off paradise will cost, then you will get expert advice to assess the cost; you will not make a blind guess. You will also feel reassured if you check the expert's opinion. This is common sense.

We are, in fact, all estimators, and a large part of what we estimate is the cost of the things we do. And this cost is the cost of providing the resources we need to do them.

The principles of estimating which we use in our domestic world are the same as those which business uses. Business success is dependent on setting a budget which we can work to—and afford—and working to it; ensuring that we know what we need, how much of it we need, and how much it costs and ensuring that we calculate the correct total price of the resources we use. In our budget we ensure that we can afford this cost, that our income will be at least equal to the cost.

The estimating of cost follows the same process, whichever industry or environment we work in, wherever it is. It is the evaluation of the costs of the resources we use.

The cost we estimate must be as accurate as possible. Business and individual survival in an organization depends on this accuracy. If you do not know what an item costs *never* make a wild guess. Always ensure that the figures you use can be substantiated.

WHAT DO WE NEED TO PREPARE AN ESTIMATE?

We need to know what resources we will use and how much those resources cost. In construction and many other kinds of work, resources are defined as:

- labour
- plant
- materials
- sub-contractors.

These will be covered in detail in Chapter 2.

Clients define their requirements in the contract documents. From these we learn the details of what we have to provide and how much we have to provide. Materials quantities can be measured. If we know the cost of a unit of each material, then we can find the total cost of the materials.

Experienced sub-contractors will price work directly for us. If we are happy with the price and their ability to carry out the work, then we can use the prices forthwith.

The measurement of the amounts of labour and plant to do the work is a different matter. The contract documents tell us what to do but not how to do it. The decision as to how we carry out the work defines the amount of labour and plant we need and for how long we need it. You *cannot* price labour and plant without knowing how to do the job. This means that the estimator must know how the work will be carried out. Practical experience is necessary.

- You cannot estimate any cost accurately unless you have the practical experience to tell you how the work will be carried out.
- You need a full and correct definition of a client's requirements — complete and accurate contract documents.

THE REQUIREMENTS FOR PRACTICAL EXPERIENCE

Only practical experience of the work itself will tell you what resources you need and how long you need them for. Once you know this, you can estimate the price.

Practical experience will cover items such as the following.

- An awareness of the contents and the requirements of the contract documents. Do they allow the work to be sensibly carried out in the usual manner or will we be restricted in our efforts? Has the work a good degree of buildability or is it tricky and complicated? What are the stated accuracy requirements? Is the programme requirement a practical one? All such considerations affect the final price and the estimator must be aware of them. Awareness will enable us to decide how we will do the work in a way which is acceptable to the client.
- A knowledge of the location of the work. Contracts distant from the company base or in new areas are often fraught with difficulty. The working conditions are unfamiliar and resource availability may be more difficult than expected.

When working overseas, knowledge of the location is crucial. I know several examples of excellent businesses obtaining work overseas, work with which they were thoroughly familiar and totally confident when working at home. Yet each of those businesses had great difficulty in meeting their objectives, and generally lost a lot of money as well.

These companies had not made adequate allowances for the working conditions, the business practices, or the resource availability relevant to the particular location.

The principles are the same as for UK projects—the estimator analyses the rate in terms of labour, materials, plant and overheads—but there are additional problems. The problems and the solutions will be different for every country.

General issues. It may prove very difficult to obtain the required information. Failure here can be crucial.

Local law and dispute procedures will apply and may not be predictable from the conditions of contract.

Currency fluctuations will probably be at the Contractor's risk, unless it is definitely stated otherwise.

War, revolution etc. *in the country of the project* will probably be the Employer's risk, but war, revolution etc. in a neighbouring country may have an effect which is difficult to prove and may end up as a Contractor's risk.

Information on weather, ground conditions, local customs, etc. may be difficult to obtain and may be unreliable.

Inflation *in the country of the project* may be an Employer's risk (although this will be based on 'official' rather than 'actual' figures), but inflation in the home country, or the country from which the Contractor is purchasing materials, will be the Contractor's risk.

Labour-related issues. Local labour and taxation laws will apply and may include restrictions on the use of expatriate workers.

Labour may need to be recruited from anywhere in the world, with considerable difference in availability, cost, work rate, customs, language, etc.

Materials-related issues. There may be local laws which express a preference for local or associated country materials, or even a boycott of some countries' products.

Non-UK standard specifications may apply.

Quality, cost and availability vary considerably around the world and sales literature may be unreliable.

Materials which are readily available in the UK may not be available locally. For example, pricing concrete may be a matter of finding a source of aggregate, opening up a quarry and so on.

Plant-related issues. Machines, spare parts and maintenance facilities may be available locally or may have to be brought from overseas.

Access to the site will be a Contractor's problem. Civil engineering sites are generally in inaccessible places!

Overheads-related issues. Calculations for site overheads may be difficult. Most overseas employers do not like the thought of paying a contribution for Head Office overheads and profit in the home country.

Allowance for local agent's fees and other factors has to be built in somewhere.

Other influential factors. Company capability has a substantial influence on success. Many excellent smaller companies struggle on a larger job. Similarly, companies which are first class when doing very large contracts can be patchy performers on smaller work. You need to know your strengths and work to them.

Ensuring that appropriate, capable resources will be available at the right place and at the right time is a fundamental requirement.

When the work is to be carried out is an influential factor. The seasonal effect on work can be considerable. River work in winter, roadworks during peak holiday traffic flows, obtaining appropriate resources in boom times, all require careful judgement.

In essence, the practical experience of the estimator must cover the full requirements of the work. Any compromise will affect the business to varying degrees and is likely to have a negative impact.

THE CONTRACT DOCUMENTS

Many of the disputes referred to arbitration are caused by faulty contract documentation. It is important that the parties to the contract, the Client and the proposed Contractor, are aware of this from the outset and that each takes the necessary actions to prevent future problems.

I feel it is wise for clients to work to standard conditions and documents as far as possible. The proposed Contractor needs to ensure that his estimating team receives commercial advice on any deviations from the standard documents which may exist.

It is preferable for the detailed design and documentation to be completed *before* tenders are sought. This will reduce the likelihood of claims by contractors and is beneficial to both parties to the contract.

Inadequate documentation usually gives rise to problems and client teams may find themselves in default of the law as a result. I refer to the duties imposed by the *CDM Regulations 1994* in this respect.

The contract documents will usually include the items detailed below.

The Conditions of Contract

- *The Institution of Civil Engineers Conditions of Contract, 6th* or *7th Edition*
- *The NEC's, The Engineering and Construction Contract*

These are examples from a number of standard contract conditions which may be used for construction work. Standard conditions define the duties and responsibilities of the parties to the contract. They allocate responsibility for the risk inherent in the work to the various parties. They detail procedures for dealing with payments, variations, disputes and hand-over of work.

From a Client's point of view, it is sensible to remember that such Conditions are prepared and agreed over a long period of time by practitioners eminent in their field. Is it really necessary, or sensible, to modify them?

From an estimating point of view, it is important to know if the conditions are standard. An estimator knows the risk implied by standard conditions and will be accustomed to pricing this risk. Changes in the Conditions will vary the normal risk allocation and this will probably affect price.

Wise organizations check the Conditions as a matter of routine. Change can affect insurance, alter the pricing strategy, or even lead to a refusal to tender.

The Specification

The Specification sets out the standards that have to be worked to, including the particular demands of the Client for various provisions on site and the programme of work.

Specifications cover items such as:

- quality of work—the tolerance to be applied during the construction work. Standard tolerances are sensibly achievable, tighter tolerance can greatly increase cost, slow down the work, and show no apparent benefit
- material strength and quality—specified to a British, or other, Standard or strength (crushing strengths of concrete, type of brick, thickness of mortar joints are examples)
- the client's requirements on site (what the Engineer is usually provided with—accommodation, clothing, instruments)
- programme of works.

In the UK there is usually little problem with materials compliance or with a sensible specification for workmanship.

The bill of quantities

The most commonly used contracts in construction are of the admeasurement, or measure and value, type. This involves the production of a bill of quantities which should sensibly measure all items of work and present them in a list.

Where a bill of quantities is not provided for those tendering for work, in 'design and construct' work for example, then those estimating must ensure that a full and correct bill of quantities is produced by their own organization. It is extremely difficult, if not impossible, *to fully and correctly price any work which is not fully and correctly billed*.

Not only must a bill of quantities reflect each type of work activity necessary to construct the works, it must also give an accurate measure of each activity.

It is sensible to produce bills of quantities in a standard manner and to standard rules. Those using them, including the estimator, then understand what is deemed to be included in the various items of the bill. This is a standard method of measurement. In civil engineering we have the *Civil Engineering Standard Method of Measurement* (currently *CESMM 3*).

In the *ICE Conditions*, Clause 57 deems that bills of quantities are prepared in accordance with the *CESMM 3* unless stated otherwise. While the *CESMM 3* covers simple building work, it does *not* cover complicated building or electrical and mechanical work. If you *do* use another method of measurement you must say so.

Any error in a bill of quantities is likely to be one of omission. This immediately leads to an underpricing. Whoever provides the bill of quantities should bear this in mind and make an allowance. I have found that an overall allowance of 3% to the net price has been adequate for this correction.

The drawings

No work can be *fully* priced without first having a *full* set of drawings. Not only do the drawings indicate what is to be done, their detailing will show if construction will be speedy and practical, or if it will be difficult and time-consuming. This, the degree of buildability, is a crucial determinant of cost. Buildability needs to be carefully considered when preparing any construction programme or the price for the construction itself.

Is the reinforcement sensibly fixable and are the bar diameters sensibly high? Is there a large number of bar marks or a small number? Is the dimensioning of the structure repetitive, or is there constant change? Do walls meet at right angles or is everything on a skew? Each of these questions, and many similar considerations, will have an effect on price which can only be judged by a careful examination of the drawings.

Some simple examples come to mind.

- A small RC pumping station seemed to be making very slow progress so I went to look at it. The reinforcement consisted of many small, short bars of small diameter — slow and difficult to fix, and hard to maintain in position. This was quite apparent when the drawings were inspected.
- A high quality concrete frame had skewed centre lines, not at 90° to each other. The extra work necessary to set out the structure and to cut the various materials used to the correct skew angle was considerable. So was the cost. When completed, however, the structure looked very like a normal rectangular construction.
- An RC frame building had brick infill panels beneath the windows on each floor. The bricks were a special handmade product and the mortar joint was tightly specified. The jointed brickwork did not fit the gap between the concrete columns, or between the floor slab and the underside of the window frame, without a modification which the architect resisted.

Each of these problems was apparent to people who had experienced the problem before and had examined the information provided.

This reinforces the point made above: estimators must have practical experience of the type of work and full information to price work correctly.

Site investigation reports

It is unlikely that a client will conform to the duties imposed by the regulations unless very careful consideration is given to a site investigation before any design or other work commences.

Information on ground strength and make-up is vital for decisions on methods of construction. Evidence of any contamination should be provided so that the necessary treatments or actions can be preplanned. Adjacent properties could well have an effect. Many modern buildings have deeper foundations than their adjacent, older neighbours. How should the new foundations be constructed when this is so?

Examination of the site investigation report will help an estimator to assess methods of excavation, contamination treatment, the needs of temporary works, the use of piling equipment and other similar considerations. Operations involving construction risk should appear in the pre-tender health and safety plan.

The information provided is an important consideration in the pricing exercise.

The pre-tender health and safety plan

The *Construction (Design and Management) Regulations 1994* require the preparation of a pre-tender health and safety plan. The intention is that those tendering for construction work can take account of the contents as they prepare their tenders.

The contents of the pre-tender health and safety plan vary from job to job and depend on the nature of the project. Areas to be considered for inclusion in the plan include:

- the nature of the project itself — its location, the type of work, the Client, the programme
- the existing environment — the surrounding land use, ground conditions, contamination, location of services and any other factors currently existing which may affect health and safety
- existing drawings and any other information on the site itself, adjacent properties, etc.
- the design itself and information on any significant risks to health and safety which remain in the construction process when working to the design
- construction materials to be used and the health hazards which arise from their use
- site-wide elements — the factors which affect all aspects of the site: access and egress, any welfare areas, lay-down areas, people or traffic routes
- overlaps with the Client's work — the risks which may arise on premises where work is to be carried out and which are occupied by the Client
- site rules — emergency procedures, evacuations, permits to work: the rules set down by the Client to ensure safe working on an in-use site

- continuing liaison—the procedures for considering health and safety during the construction phase and the arrangements to deal with any problems which may arise.

Clearly, the contents of the pre-tender health and safety plan can have a considerable effect on price. Examples of problems which I have had in the past which could have been mitigated by earlier consideration and discussion include the following.

- The basement of a new structure was some 3 m deeper than that of the existing structure immediately adjacent to it. There was no room to provide temporary works to facilitate excavation or construction of the new basement along the side common with the existing structure. A slightly smaller new basement would have provided space for temporary works along the line common with the existing basement.
- Work on deep structures in a chemical plant was seriously delayed by the need to check the area for gases (like a confined space) prior to commencement of work. Checking requirements elsewhere by the Client meant that work start was delayed until mid-morning. In addition, the number of workers in the structure at any given time was seriously restricted and lifting plant was duplicated. It is vital that such requirements are made apparent, and are allowed for, in any tender.
- The temporary works on a river scheme involved the use of cofferdams. To prevent river flooding, the tops of piles in the cofferdams had a specified height of 300 mm above the specified normal water level. As construction work progressed it was apparent that the 'normal' water level was higher than indicated. The works had less than the 300 mm freeboard intended. Flooding risk was increased.
- The refurbishment of a historic steel structure involved the removing of the original paintwork and its replacement with an expensive modern product which had to be applied at a high ambient temperature. The work was carried out in winter and it was necessary to provide a heated temporary cover over the works. The original paint had a very high lead content, the new produced unhealthy fumes. Extreme measures of people protection were necessary.

I hasten to add that the examples given occurred prior to the *CDM Regulations* being introduced. Problems were resolved and work was completed successfully. The examples do, nonetheless, give an indication of what can happen and what we need to be aware of either to prevent the problem in the first place, or to allow for the provision of necessary measures to overcome the problem as we prepare our estimates.

SENSIBLE TENDER PERIODS

As stated above, construction estimates are prepared by pricing the resources used, those resources being:

- labour
- plant
- materials
- sub-contractors.

While these will be discussed in more detail in Chapter 2, it is *largely* correct to say that, in estimating UK construction:

- labour costs are standard in an area and referenced to the *Working Rule Agreement*
- plant is readily available and competition keeps prices at a sensible, predictable level.

Estimates of the labour and plant elements therefore tend to have standard rates in the estimating process.

Prices for the materials and sub-contractor resources are *not* standard, however, and good working practice demands that fresh quotes for these items are sought for each tender.

In addition, suppliers must price items, or work, according to the same conditions of contract to which the tenderer is working. Suppliers have to be provided with the relevant items of the Specification, Drawings and other data from the contract documents. The copying of this data takes time. It can take three days of copying to provide sufficient data to enable all prospective suppliers to price their offers in full compliance with the requirements. Priced offers from prospective suppliers are received some two weeks later and the selected offers can then be included in the price preparation exercise.

A medium-sized contractor, with a turnover of around £50 million, is likely to have a tendering success ratio of about 1 in 5. For each tender success, five tenders have to be prepared. With an average tender price of £500 000, the annual requirement on the estimating department is to produce 500 tenders with a total value of £250 million — an average of eleven tenders per week.

Such a tender output can only be sustained by a structured estimating process. A conveyor belt system is needed. Each estimator will be working on a number of estimates at any given time.

Accepting that necessary pricing information for any tender bid will not normally be returned before the end of the third week of the tender period, the sensible logic is to allow a four-week tender period as the norm for 'construct-only' tenders.

Where 'design and construct' or similar bids are requested by clients, the time for design, investigation and quantifying needs to be allowed. The time needed for this clearly varies from job to job. A minimum tender period of around twelve weeks for a job of any reasonable size seems a sensible guide.

The crucial point to remember when setting an appropriate tender period is that it is in everyone's interest to allow reasonably sufficient time. Failure to do so may well put organizations in breach of the regulations.

Moreover, it takes time to prepare a sensible estimate. Sensible estimates are less likely to give rise to problems at a later date. An adequate tender period makes commercial sense to all the parties concerned.

PREPARING PRICES — THE KEY POINTS

Construction can usually be divided into Building or Civil Engineering work.

Building work

Large elements of building work, on any contract in any area, have a predictable cost. The prices for internal finishing work — all timber and joinery, blockwork, plastering, painting, decorating are examples — vary little between sub-contractors and between contractors. Prices for such work in different areas will change, but the change will be area dependent and not work dependent. The likely exception to the rule is where there is a large demand on a particular trade (bricklayers and plasterers are a case in point) and this demand can lead to a rapid price rise.

By and large, however, sub-contractors will submit prices in a given area which will compare closely item for item, job by job. The reason for this is that the work is predictable and unlikely to change in content or complexity. It is standard work carried out under standard conditions.

As prices for such work vary so little they form what is virtually a schedule of rates or a standard price-list. In the absence of a quotation, the standard rates for an area can often be used without too much risk.

Prices for exposed work — brickwork, concrete, etc. do vary from job to job as the job itself changes, but this variation is usually relatively small.

The traditional pattern of using sub-contractors for most builder's work packages and the regular nature of the prices themselves leads to contractors often submitting almost identical prices for large areas of any proposed work.

Tender prices for building work, certainly if it is of a traditional nature, tend to be relatively close to each other. The fact that there is little risk means that 'standard' prices need little alteration. The difference between the highest and lowest bids on a new £10 000 000 retail store is likely to be in the order of as little as £150 000.

The conclusion we draw is that, for many areas of building work, standard schedules of prices can be used. However, a sensible estimator will always ensure that any rate is relevant to the item being priced before using that rate.

Civil engineering work

While schedules of rates (or prices) can be used to price many areas of building work, to do the same for civil engineering works can be bad practice and should be avoided (unless you are quite certain that you are correct). The reason for this is that the nature of civil engineering work is much more varied.

River works can be much more expensive to carry out in winter than in summer.

The cost of a pipe in a trench will vary with location, trench depth, type of ground and many other factors. The cost will vary dramatically between work in roads, fields, gardens or footpaths.

The cost of excavation will vary according to the type of ground, the depth of dig, the distance to tip, the location of the site and many other factors.

The cost of placing concrete varies depending on whether it is in thick slabs, in walls or columns, on the ground or in the air, and whether it is being placed in large or small volumes.

Reinforcement is cheaper to fix in large diameters than in small ones, and if there are only a small number of bar shapes (and fixing is repetitious).

It is usually unwise and often foolish to price such work on a schedule of rates basis. What you need—and this is the basis of civil engineering estimating—is:

- knowledge of exactly what you have to construct
- practical experience of the type of work you are pricing and how the work is carried out
- knowledge of the resources you need to carry out the work
- awareness of the time it takes to complete the various items of work
- knowledge of the cost of each resource.

Once you have this information, the cost of any item can be calculated. It is simple arithmetic. It is *estimating from first principles*.

Some clients prepare budget prices for construction work *using schedules of rates*. We will discuss schedules in more detail in Chapter 5. While this practice will give a guide to overall price, it should not be regarded as an accurate yardstick for the cost of each element of the work. Other clients request tenders, usually for work of a minor nature, which are based on a schedule of rates. There is usually no guarantee of continuity of the work, or the location or amount of work at any one time. As a result, tender prices can be high. As it is usually the lowest price which is successful, this is likely to be too low and based on optimistic assumptions.

You should accept that schedules should be used with care in civil engineering, and perhaps avoided.

LARGE JOBS, SMALL JOBS — BITE-SIZED BITES

I have heard experienced engineers make statements such as "a kilometre of motorway costs £10 000 000". This sounds very impressive —but is it?

The fact is that the cost of a motorway depends on many factors such as:

- is it urban or rural?
- how many traffic lanes are there?
- are the structural works extensive?
- do the earthworks quantities balance?
- are earthworks extensive?
- what is the ground like?

These and many other factors will determine how much the kilometre of motorway *really* costs. They make a nonsense of the bland overall statements we have all heard. Do not fall into the trap of making such errors yourself.

How do we price large contracts? Inexperienced engineers will insist that they are quite incapable of such work and do not know where to start. The probable truth is that *no one* can price large jobs as a *whole*. We all split large jobs into smaller jobs and smaller jobs into their respective elements. We can then usually price the elements.

You split large jobs into smaller jobs and then into '*bite-size bites*', pieces of work that are small enough for *you* to price. For example, a road contract would split down into:

- earthworks
- drainage
- carriageway
- fencing
- signs
- bridges, etc.

Each bridge would be split into:

- excavation
- piling
- foundations
- abutments
- wing walls
- bridge deck
- surfacing
- footpaths
- fencing, etc.

Always remember the point about 'bite-sized bites'. If you *do*, you are likely to be able to price work better than you expected to; if you *don't* then you are likely to fail.

As you work out your prices for the 'bite-sized bites' you add them together and the total gives you the price of the motorway. All tender prices are built up in this way. The bill of quantities fortunately reflects this approach and is an excellent guide.

Do not 'guesstimate' the price of work by throwing a sum of money at it, work it out little by little and understand what you are doing.

THE DIFFERENT TYPES OF COST

It is important that you understand the different types of cost which businesses use and refer to on a regular basis. You also need to know how estimators use these costs. The types of cost we refer to are detailed below.

Quoted costs

As explained earlier in the chapter, quotations are sought for materials and sub-contractors, while labour and plant costs are the standard costs used within the particular business. Priced quotations are returned by prospective suppliers and the rates which they quote are the *quoted costs*. The most attractive supplier rates are used by estimators to calculate the appropriate rates to be used in the tender.

Net costs

The costs actually used to prepare the tender are the net costs of the work. These are the quoted costs plus allowances for waste, handling and any other expenditure necessary to carry out the work. There is no allowance for inflation and costs are *current* ones. Inflation adjustment will be considered below.

Net cost is the actual sum of money expended to carry out the work. Estimators price work initially in terms of net cost. For example:

> Net cost of reinforcement = quoted cost of reinforcement
> +
> waste
> +
> tying wire
> +
> unloading and placing
> +
> fixing
> +
> any cranage used

All materials have an element of waste which varies from material to material and location to location. This will be explained in Chapter 4.

Remember that all work is initially priced on the basis of net cost

Inflation cost

If we consider the time-scale of any project we will realize that inflation will affect all of the costs which we use. Inflation will affect some parts of the work (often the later items) more than others (often the earlier items). How should we reflect this in the price?

As the inflationary effect is continuous it will be very time-consuming to calculate its effect on each item of work. The sensible approach is to price all work on the basis of current net prices. When this exercise is completed the inflationary effect is considered in a single calculation which covers the whole of the work. We will cover this in Chapter 2.

Overheads and profit

Every business has an *overhead*, the element of cost which cannot be allocated to the direct operations of the business. Overheads cover items such as:

- directors
- secretariat
- buildings
- accounts
- marketing
- furnishings
- equipment, etc.

The overhead is expressed as a percentage of the turnover of the business and added to each tender as a percentage of the net cost.

As the purpose of business is to make a *profit*, it is usual to add a further percentage to the net cost to allow for this.

Total cost

The total cost of any item = net current cost
+
inflation
+
overhead
+
profit

This is the sum which the contractor puts into a tender. It is intended to:

- allow for the payment of all costs including inflation
- cover the cost of the company overhead
- enable a profit to be made.

Grouping items for pricing purposes

It is essential business practice that contractors submit the correct tender price. It is not so vital that every rate which a contractor puts into the bill of quantities is correct. Of course, rates must be reasonably correct, but it would take far too long to calculate every rate exactly and the time available is better spent on wider strategic considerations — overall price, needs of the business and resource availability are some of these.

Also, the estimating process is an art rather than a science. The estimator's opinion will determine many of the prices and 'opinion' is exactly what is involved.

What we actually do is to examine the bill of quantities and assemble the quantities for the various tasks which are to be carried out. Excavation, concreting, reinforcement fixing, formwork are often to be carried out in several locations. We look at each task group of items and, as far as possible, price them globally. We then split the global price into respective items, probably weighting each type of work for complexity.

Let us take some examples of what the estimator is likely to do to simplify the task. We will assume that the bill of quantities — since we cannot sensibly price work without one — is available and has been prepared in accordance with *CESMM 3*.

Provide concrete. The price quoted by concrete suppliers depends on the concrete mix being supplied. Contracts often require several mixes, each of which is likely to have a different price.

The amount of concrete wastage varies depending on where the concrete is placed. Blinding, concrete in trenches, thick slabs, thin slabs, walls and columns all tend to have significantly different wastage factors. These factors need to be considered as we consider the appropriate rate to put into the bill of quantities.

Place concrete. Larger contracts tend to have a full-time concrete gang which has its own allocated plant. The estimator works out the total cost of concrete placing and spreads the required total sum across the various concrete placing rates.

Placement in large thick slabs will be relatively cheap. Placement in columns will be expensive.

Supply and fix reinforcement. Reinforcement fixers tend to be self-employed sub-contractors who price the reinforcement as it is billed, by the tonne, and do not differentiate the different areas by price.

18

Large diameter reinforcement is cheaper to fix per tonne than small diameter (since there is less length to fix).

The rate for supply is fixed nationally for each diameter and extras such as wastage are taken as a standard percentage. The result of this is that reinforcement rates on a contract tend to be standard prices per tonne of each diameter, regardless of where the steel is fixed in the structure.

In structures such as water treatment plants there is a requirement to fit many pipes of differing diameters into concrete walls. For simplicity it is common practice to calculate the cost of fitting the largest and smallest pipe diameters and to price the other diameters on a pro-rata basis. The same practice would apply to the fixing of pipe fittings. The material cost will, of course, always be that quoted.

Much of the work in a tender is priced by using sub-contractor rates directly from a quotation. This helps estimators a great deal, saving the substantial time it would take to calculate the rates themselves.

An important point to note when pricing groups of items is this:

- it is *usually reasonably accurate* to price the largest and smallest items in any group and to establish rates for intermediate items on a pro-rata basis
- it is *often inaccurate* to price two items with adjacent sizes and establish rates pro rata up or down the scale for sizes above or below those priced.

THE PITFALLS TO AVOID

For any pricing exercise to be carried out sensibly you should observe the following criteria.

- You need practical experience of a type of work in order to price it. If you *do not* have the relevant experience and need to price some work, discuss the requirements with someone who *does* have the necessary practical experience.
- Never panic. Think the problem through calmly, seek help as necessary. You *can* do it.
- Never make a wild guess at a price — always work from first principles.
- Consider every price as being built-up using the constituent elements of labour, plant, materials, sub-contractors.
- Never try to price an element of the work which is too big or too complex for you to understand. Split it down into smaller pieces. Go for 'bite-sized bites' that suit *you*.

- When you study the contract documents, look for the following points.
  - Are the contract documents complete? The less complete the data you are given, the more likely you are to have problems.
  - Are the Conditions of Contract standard? Changes tend to increase the risk borne by the contractor.
  - Is the Specification a standard one? Change, of tolerances for example, is likely to *increase* construction costs.
  - Are the drawings complete? What is the level of 'buildability'? High buildability reduces cost and programme time. A low degree of buildability will have the opposite effect.
  - Does the bill of quantities look adequate? If *you* take the risk on quantities it *must* be adequate.
- Remember that estimating departments have a continuous throughput of tenders and that it is therefore difficult to treat any tender as a 'special case'. Ensure that you allow adequate time for any tender to be prepared.

This chapter has offered some general guidance on the estimating process. Chapter 2 will consider the elements which are used to build up the prices in the tender.

2

Building up the net price

Every construction company has its own method of evaluating the tenders which it submits. The methods will differ between competitors in the industry. Provided the exercise is carried out correctly, however, the differences can be seen to be cosmetic rather than fundamental.

The estimator traditionally calculates the net prices required to carry out the work. These are put into the bill of quantities after a percentage for overhead and profit has been added by a senior manager. The net prices in the bill are based on the net cost of the resources we use in any bill item. The net tender total will be the aggregate total of all net costs. You will understand this more readily when we price some examples in later chapters.

Estimators must have practical experience of the work they are pricing. They need to know how the of work will be carried out, what resources will be used and how long it will take, or how much resource will be used. If estimators know the cost of the various resources they require, then the net price of items can be estimated.

The price of any work depends on factors such as the following.

- How much labour do we need and how long do we need it for?
- What plant do we need and how long do we need it for?
- How much material do we need and what should we allow for waste and other factors?
- What sub-contract resources will we use?

These are questions which are asked in every tender by every estimator and it is the answers to them which determine the net price.

Each estimator preparing an estimate will have a different opinion of what resources will be used and for how long. Some resource costs vary from one company to another. So rates will vary between companies and different tender totals will be submitted. If the work being priced is routine and standard practice, with little risk, the contractors' tenders are usually close together. If, on the other hand, conditions are onerous and the work risky, then there is likely to be a wide divergence in price.

We will now consider how the net price is built up.

Net prices for work are derived by calculating the net cost of the resources used to carry out that work. In construction the traditional elements of resource are:

- labour
- plant
- materials
- sub-contractors.

Companies in the UK tend to use:

- standard prices for labour which are revised when industry rates of pay are changed
- standard rates for plant — there is a readily available supply of good plant.

Special items will be subject to a quotation:

- we need the appropriate prices for the materials we will use
- we need quotations from sub-contractors for specialist services.

You can see that we must get prices from various suppliers for materials and sub-contract services. We request prices by sending out *enquiries*.

ENQUIRIES

Contractors agreeing to work for Clients agree to work to the conditions of contract, the specification, etc. in a legally binding manner. The contractor needs to ensure that suppliers of all kinds conform to the same conditions as the contractor does in the main contract to the client. Enquiries are therefore sent out which include relevant parts of the specification, drawings, bill of quantities, conditions of contract etc. Suppliers are expected to conform to the conditions of the main contract. If their default causes the main contractor to default, then action can be taken by the contractor against the defaulter.

Quotations are received from the various suppliers in response to the enquiries and these are checked to ensure that they comply with the necessary conditions, are arithmetically correct, and that all items are priced.

Favoured quotations, usually those offering the best service at the lowest price, are put forward for inclusion as resource prices in the tender rates as the rates themselves are prepared.

BUILDING UP THE NET PRICES

As prices vary from one organization to another, I have used data from the *CESMM 3 Price Database 1999/2000* edited by E.C. Harris and published by Thomas Telford Publishing. The database gives data collected from many sources on material prices, waste factors, labour and plant outputs. It covers all items of the *CESMM 3* bill of quantities in an easily understable method. Prices in the database are reasonable. In competitive situations they are likely to decrease.

Having identified the four resources, we will now see how net prices are calculated for them.

Labour

Here we refer to operatives and tradesmen directly employed by a company. Companies use their own rates for labour which are usually based on the *Working Rule Agreement* published by the Construction Industry Joint Council. Council members are employers or trade unionists who work within the industry. The Council has an ongoing role in keeping wage rates and working conditions at acceptable levels. Agreements are updated, traditionally on an annual basis, and set out the current agreements on rates of pay and conditions. The agreed rates and conditions are the *minimum* ones. Employers can pay more than the agreed rates, but not less.

The *Working Rule Agreement* covers conditions such as:

WR1 basic and additional rates of pay
WR2 bonus payments
WR3 working hours, rest and meal breaks
WR4 overtime rates
WR5 daily fare and travel allowances.

The full Agreement will not be covered here but it is recommended that all those in the industry be aware of it and its contents. The appropriate rate for labour to be used in any pricing exercise is derived from it. Table 1 gives a typical example showing how labour rates are calculated. This is a *minimum* rate. In practice, somewhat higher rates are likely to apply. Estimators use the relevant local rate.

Table 1. *Typical labour costs—based on a five-day week (45 working hours)*

White Rose Construction				Labour costs from 28 June 1999		
		Craftsmen			Operative Grade 3	
Based on 9 hour day, Monday to Friday, 45 hour week	Basic rate	£	p	Basic rate	£	p
Basic wage—5 days × 9 hours = 45 hours worked + 3 hours non-production 48 hours at basic rate	6.05	290	40	5.20	249	60
ADD Attraction bonus		20	00		15	00
Total taxable pay		310	40		264	60
ADD National Insurance (10%)		31	04		26	46
Holidays with pay		21	30		21	30
Public holidays		8	33		7	16
(64 hours paid in 46½ weeks worked)						
Sick pay		2	00		2	00
Redundancy and training (7½% of wages)		23	28		19	85
Guaranteed minimum (5% wages)		15	52		13	23
Total for 45 hours		411	87		354	60
Average hourly cost		9	15		7	88
To be used in all tenders		9	25		8	00

On this basis, the rates to be used in tenders as at 28 June 1999 are:

For tradesmen   £9.25 per hour
For operatives   £8.00 per hour

The above rates will be used in the examples in later chapters. They *exclude* allowances for the following.

- *Travel time* to work, paid as a daily sum based on the distance travelled. This clearly varies from job to job. A general allowance can be added to the basic rate and used on every job, or alternatively a specific allowance can be added to the tender for each job, dependent on location. My experience has been to
  ○ use the basic rate for all cost calculations

○ add a specific travel allowance in the site set-up costs or preliminaries (covered in Chapter 3).

This practice will be applied in later examples.

- *Subsistence.* Distant contracts require people to work away from home and subsistence payments are made to cover the costs. Such allowances are commonly covered in the preliminaries (Chapter 3).
- *Tea breaks.* The morning and any afternoon tea breaks are paid as working time. Lunch breaks are not paid for. I have never known labour rates to be enhanced to cover the cost of tea breaks. Normal working hours give rise to a morning break but not an afternoon one. A sensible increase in the labour rate to cover this would be $7^{1}/_{2}\%$. This will be discounted in future examples.
- *Down time.* Contracts to be carried out in exposed conditions can suffer severely due to the effect of weather. Any 'down time', time lost due to conditions, should be taken into account on those contracts affected. This is usually a 'one off' calculation, again incorporated within the preliminaries.

Working on the basis outlined above, the same basic rates for labour can be used in each tender, and suitable additions made for the extra labour costs appropriate to a particular job are also included in a 'one off' calculation in the preliminaries.

Plant

Plant can be owned by the company or hired from external sources to suit the needs of a particular job. As stated earlier in this chapter, competition has made plant hire rates fairly standard. A company with its own plant fleet will sensibly charge market prices for hiring out.

The net price of plant is built up from a number of items as given below.

- The hire charge itself. This is standard for the item in question.
- Transport on/off site. All plant, with the exclusion of mobile cranes or JCB type wheeled machines, has to be transported to and from site, incurring a charge.
- The cost of fuel. The amount of use to which plant is subject varies from site to site. Experienced plant managers will give a good guide and you can add standard allowances without too much risk.
- Extra equipment. The hire rates quoted for items such as compressors and water pumps will be the rates for the basic equipment only. For a compressor you are likely to need extra hoses and a variety of air tools. For a water pump there will be extra hoses

for both suction and delivery. These extras are likely to double the basic hire rate and should not be ignored.

- Set-up costs. Some items of plant require special facilities:
  o heavy duty electric pumps will require a three-phase electrical supply
  o tower cranes may need heavy foundations, which may also have to be removed once construction is completed.
  o a heavy mobile or crawler crane may require a heavy hardstanding
  o temporary works, such as bridges, cofferdams, etc., may be required
  o timber mats may have to be supplied for crawler cranes
  Due allowance needs to be made as these occur.
- Operators. Plant items such as cranes and excavators are supplied with an operator and this has to be covered in the rates. The basic rate for the machine usually includes the operator, but extra charges could well accrue for the operator's
  o overtime
  o maintenance
  o travelling time
  o subsistence.

Table 2 gives examples of typical plant costs and how they are derived. These costs will be applied in the examples in later chapters. The rates given are sensible ones. Competition may well lead you to use lower rates than these.

Materials

The *enquiries* sent out for supplies of materials for the contract for which we are tendering result in *quotations* from suppliers who are able to provide the specified items. Quotations are checked for:

- compliance with the terms of the contract
- adequacy of supply (the supplier can provide what we want, when we want it)
- price.

The most favourable quote for each required material—usually the cheapest offer made by a supplier—is then used in the build-up of material prices.

In addition to the cost of the material itself we need to allow for:

- offloading

Table 2. *Typical plant costs (in £)*

Plant item	Weekly hire	Fuel	Consumables	Operator	Total weekly cost
40 t crawler crane	550.00	25.00	32.00	373.00	980.00
30 t crawler crane	450.00	15.00	27.00	373.00	865.00
25 t mobile crane	430.00	345.00	26.00	373.00	1174.00
180 cfm compressor	90.00	32.00	9.00		131.00
Compressor tools	30.00				30.00
1.5 t dumper	70.00	17.00	3.00		90.00
Power float	55.00	10.00			65.00
Fork lift 2.0 t	200.00	30.00	5.00	355.00	590.00
JCB 3C	250.00	30.00	7.00	373.00	660.00
Komatsu excavator	500.00	67.00		373.00	940.00
Kango drills	20.00				20.00
5/3$^{1}/_{2}$ mixer	30.00	15.00			45.00
Flygt 2" pump	40.00	10.00	10.00		60.00
Flygt 4" pump	65.00	25.00	10.00		100.00
Flygt 6" pump	140.00	100.00	60.00		300.00
75 m diesel pump	65.00	5.00	15.00		85.00
Cat 951	325.00	55.00	17.00	373.00	770.00
Transit	130.00	80.00			210.00
Escort	115.00	60.00			175.00
Land Rover	140.00	80.00			220.00
Minibus	130.00	80.00			210.00
1.3 saloon	95.00	30.00			125.00
1.5 saloon	110.00	35.00			145.00
2.1 saloon	125.00	45.00			170.00
2.0 Ghia	145.00	45.00			190.00
24 t tipper lorry	500.00	250.00		355.00	1105.00

- distribution on site
- waste, etc.

Offloading and distribution on site are usually included as part of the site set-up or preliminaries cost which will be considered in Chapter 3. We do, however, need to allow for wastage at this stage.

Table 3 gives some typical costs of materials which will be used in later examples. However, competition could well reduce these prices.

Sub-contractors

All businesses—and the national and international economies are businesses—work to the appropriate business cycle. They experience both high (business peaks) and low (business troughs) points. The high and low points are those of the industry or the national economy. There is little a business can do except to manage *itself* to conform to the existing cycle.

Figure 1 shows such a business cycle.

- Point B is the point of greatest activity in the cycle. If our business

27

Table 3.   Typical materials prices (in £)

Material	Base cost	Wastage	Extras	Total net cost
Concrete C15/20 mm	52.68	30%		68.48/m^3
Concrete C30/20 mm	58.73	5%		61.67/m^3
Aggregate				
40 mm limestone	6.89	10%		7.58/t
Zone 2 sand	7.25	10%		7.98/t
18 mm plywood	5.30	15%		6.10/m^2
Cut, bent, bundled				
MS reinforcement 10 mm	295.00	5%		310.00/t
MS reinforcement 20 mm	273.00	5%		287.00/t
MS reinforcement 25 mm	273.00	5%		287.00/t
MS reinforcement 32 mm	285.00	5%		299.00/t
MS reinforcement 40 mm	287.00	5%		301.00/t
PAR timber, wrot	375.00	15%		431.00/m^3
A393 mesh fabric	1.85	10%		2.03/m^2
Clay pipes to BS65				
150 mm dia.	8.93	5%	0.37 unload	9.75/m
300 mm dia.	29.77	5%	1.27 unload	32.53/m
375 mm dia.	51.36	5%	1.69 unload	55.62/m
Flexible concrete pipes				
BS5911 class M				
600 mm dia.	27.10	5%	1.20 unload	29.66/m
750 mm dia.	45.75	5%	1.51 unload	49.55/m
1200 mm dia.	106.65	5%	1.99 unload	113.97/m
Brickwork, etc.				
Class B engineering	95.01	3.5%		98.34/1000
100 mm concrete blocks	3.86	3%		3.98/m^2
100 mm artificial stone	16.42	5%		17.24/m^2
Cotswold rubble 100–200 mm	32.64	10%		35.90/m^2
Mortar, gauged, coloured	80.00	5%		84.00/m^3

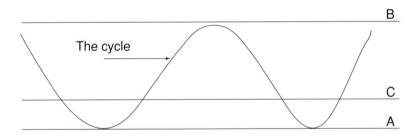

Fig. 1.   The business cycle

A = point of lowest activity in the cycle
B = point of greatest activity in the cycle
C = employment factor in a well-managed business

employs sufficient numbers of directly employed people to cope with work at this level of activity we will be able to handle the maximum amount of work which we are likely to get at any time. As the volume of work decreases from the peak, as it inevitably will, we will be in an increasingly difficult position as increasing numbers of our employees will be short of work. We will be paying the same peak costs to produce less work. This is an inefficient way to run a business and is likely to end in bankruptcy. The traditional approach, of hiring and firing people to enable the business to follow the pattern of the cycle as closely as possible, is no longer acceptable. It is a sign of failure. Employment at level B is not regarded as an option in the construction industry.

- Point A is the point in the cycle with least activity. If we directly employ people to cope only with this level of business, then our risk will be minimized. There will, however, be no slack in the organization to allow for holidays, training or any of the activities which encourage people to perform better. Everyone will be working flat out at the lowest level of business activity. The slightest sign of an increase in the level of activity will see us struggling to cope.

- Point C is the sensible level of employment. It needs to be as close to point A as is reasonably possible. Our business can then work as a business should work and we can be professionals in all senses of the word. Business activity at levels above point C will be covered by sub-contracting.

While this is a simplistic explanation, the fact remains that sensible businesses need to employ sub-contractors. The key is to employ them effectively! Sub-contractors help us to spread the risk inherent in the work. Commercial pressure makes them a necessity.

The sub-contract price to us is a series of individual rates for the relevant items of the bill of quantities. However, these rates are calculated by the sub-contractor as elements of:

(labour + plant + materials) + overhead and profit

The enquiry sent to each sub-contractor contained the necessary documentation to provide a full brief on the work scope.

- The drawings indicated *what* they had to do.
- The bill of quantities showed how *much* they had to do.
- The specification defined the *standards* to which they had to work.

Each sub-contractor would have been informed of any health and safety risks inherent in the job, and the sundry programme requirements. They should therefore know as much about the job as we do and be able to supply all-inclusive prices. We now take each area of the work where sub-contract prices were sought and compare the prices received to ascertain the most attractive offer.

To find the most attractive offer we need to ensure:

- that each sub-contractor prices all the items
- that there are no arithmetical errors in the sub-contract totals
- that they can comply with our construction programme and supply sufficient competent resources to keep their work to the programme
- that they are reliable
- if we will be working to a fixed price, they need to do the same
- that they will comply with the same contract conditions as we do.

SUB-CONTRACTOR ATTENDANCES

Sub-contractors often need assistance (attendances) while they are on site carrying out the work, and such attendances need to be agreed and included in the tender price. Examples of attendances are:

- the provision of cranage or other lifting facilities
- the unloading of materials and equipment
- removal of the arisings from boreholes
- pumping of groundwater
- provision of welfare facilities.

Attendances tend to be evaluated and included in the site set-up costs, or preliminaries, which will be covered in Chapter 3.

Once we have decided which material suppliers and sub-contractors to use, and have taken the quoted prices and allowed any wastage, attendance or other costs, we will be in the position of knowing the *net* prices for all material supplies and sub-contract elements. We already know the rates that we will use for labour and plant.

We will then use the net prices to calculate the required rates for the construction work in the bill of quantities—that is the excavation, formwork, concrete pipework, etc. In Chapter 4 we will see how to build up our prices for this work using net prices for the four types of resource.

The site set-up, or preliminaries, the provisions to be made for site staffing, accommodation, etc. will be covered next in Chapter 3.

3

The site set-up — preliminaries

The site set-up is usually defined as the resource items provided on a site which do not form part of the requirement for the permanent works. Each site has its own requirements and these are priced as provided.

Companies often have standard checklists of the items likely to be required. Such a checklist might include:

- staff — both visiting and resident on site
- attendant labour — canteen, dumper driver, banksman, for example
- staff cars and site vans — hire, insurance, fuel, maintenance
- offices, canteens, stores and toilet facilities
  - ○ hire, transport to and from site, furnishings, cleaning, consumables, equipment
- services — telephones, mobiles, electricity, water
- site compound — hoardings, gates, fencing, hardcore areas
- health and safety — protective items, signs, clothing, training, first aid
- sub-contractor attendance — provision of labour and plant as agreed to assist sub-contractors
- labour — travel time/subsistence
- plant — plant needed on site and not already included in the resources provided for the permanent works
  - ○ extra crane visits
  - ○ additional pumping needs
  - ○ scaffolding — allow supply and erect, hire, dismantle costs
  - ○ hoists
- sundry items — small tools

- setting-out equipment—levels, staffs, theodolites, timber for profiles, paint, pins, tapes, distomats, electronic equipment
- signs—company and regulatory
- temporary works
- temporary roads
- protection of existing property, works, roads, etc.
- testing
- security—night watchman, etc.
- winter working
- temporary coverings
- site transport
- clear site—including cleaning as necessary
- insurances and bond
- inflation
- fees.

WHAT DO THESE ITEMS COST?

The actual cost varies between similar items of the same category—staff salaries or car costs are an example. It would be unrealistic to price work on this basis. Companies therefore use standard rates which are the average for their particular business. Rates will vary between one company and another and between areas. You need an appreciation of the rates relevant to your working area and your company.

Typical costs, and hence the net charges included in our tender or price evaluation, are given in the following guide. Use your own knowledge to confirm that they are acceptable in your business.

I have again used the data available in the *CESMM 3 Database 1999/2000*. The data is gathered from a wide variety of sources and is readily available. It is unlikely to show the individual variances which a particular company may have.

The rates have been rounded to comply with normal practice. While the rates are sensible and reflect true costs, you could well find that competition, especially in the provision of plant, could drive the rates lower. On the other hand, certain trades, plastering for example, are under strong pressure due to the current level of house building where plasterer rates are approaching £20.00 per man-hour.

The rates are given in Tables 4 to 11 below. All are *net costs* per week.

Staff

Table 4 details typical staff salaries. Visiting staff, the contracts manager and the quantity surveyor, for example, will be charged at a percentage rate (20% for one day per week, 40% for two days, and so on).

Table 4. Typical staff salaries

Post	Typical salary	On cost at 40%	Weekly charge (based on 45 weeks per annum)
Contracts manager	35 000	14 000	1 100
Agent	30 000	12 000	930
General foreman	28 000	11 200	870
Engineer	20 000	8 000	620
Quantity surveyor	20 000	8 000	620

Attendant labour

Attendant labour will be charged to sites at cost. The rate included in the tender will be the labour rate quoted in Chapter 2: £8.25 per man-hour.

Cars and vans (charged as plant)

Table 5. Typical vehicle costs.

Category	Hire	Basic fuel	Further fuel* (example)	Total weekly
Small car	95.00	10.00	20.00	125.00
Medium car	110.00	15.00	30.00	155.00
Large car	125.00	15.00	30.00	170.00
Small van	115.00	30.00	30.00	175.00
Large van	130.00	35.00	50.00	215.00

Van drivers to be priced as *attendant labour.
*Further fuel is taken as that used to travel to/from site on a weekly basis.

Offices, canteens etc. — including cleaning (charged as plant)

Table 6. Typical site accommodation costs.

Item	Transport each way	Hire	Consumables	Equipment	Total weekly
5 m unit	80.00	20.00	15.00	10.00	45.00
7 m unit	100.00	27.00	22.00	15.00	64.00
10 m unit	150.00	35.00	30.00	20.00	85.00
Small toilet	30.00	45.00	20.00		65.00
Large toilet	40.00	75.00	35.00		110.00

Services

Table 7. Typical site services.

Item	Sub contract install	Weekly consumables	Weekly charge	Weekly total
Telephones	150.00		45.00	45.00
Water	750.00		5.00	5.00
Mobile phone			15.00	15.000
Fax machine	150.00	5.00	20.00	25.00
Copier		20.00	20.00	40.00
Foul water	1000.00			
Electricity	500.00		50.00	50.00

The site compound

Table 8. Typical compound requirements.

Item	Supply				
	L	P	M	S/C	Total
1 × 6 m gates				250.00	250.00
100 m × 1.2 high fence				500.00	500.00
400 m² × 150 parking			1800.00	500.00	2300.00
Reinstate parking				650.00	650.00
100 m × 4.2 m hoarding	500.00	200.00	1500.00		2200.00

L—labour, P—plant, M—materials, S/C—sub-contractors

Health and safety

Table 9. Typical health and safety requirements.

	L	P	M	S/C	Total
Protect public (warnings)			100.00		100.00
General signs			100.00		100.00
Clothing—24 sets			1200.00		1200.00
Training—72 man-hours	594.00				594.00
First aid	100.00		75.00		175.00

Sub-contractor attendances—paid as labour and plant

Every contract will have its own specific sub-contractor attendance requirement. Table 10 is an example.

Table 10. Examples of attendances.

Item	L	P	Contract total
Provide welfare			Include in canteen costs
Pump arisings	50.00	100.00	150.00
Remove arisings	75.00	250.00	325.00
Provide cranage to offload supplies	70.00	150.00	220.00

Labour travel time — paid as labour

When we built-up the labour rate in Chapter 2, I explained that the cost of travel time varied with the distance travelled to work and was defined in the *Working Rule Agreement*. A typical payment, for a journey of 26 km, is £1.60 per day. This covers journeys where free transport is provided by the employer.

Table 11. Calculation of labour travel time payment.

Item — total labour in bill of quantities	Daily allowance per man	Total travel allowance
= £100 000 = 1307 man-days (based on 9 hour day at £8.50 per man-hour)	1.60	2090

Sundry plant

Most plant in civil engineering contracts is already included in the work rates, however the following occurrences are common to almost every contract.

- A lorry arrives unexpectedly from the depot with various items for the site.
  This could well occur every two weeks.
      Cost — say £140 per visit
- The site cranage will occasionally be in full use when extra lifting requirements occur.
  We hire a mobile crane to cover the need.
      Cost — say £150 per visit
- Scaffolding costs are difficult to incorporate in the work rates of the bill of quantities so we tend to put them into the preliminaries.
  The costs are in three parts:
  ○ deliver and erect — lump sum
  ○ hire charge — weekly sum
  ○ dismantle and remove — lump sum.
- If a hoist is in use on site, the cost is again difficult to put into the work items.
  The cost will be built up in the same way as the scaffold cost.
- Additional pumping needs.
  Excessive demands, for dewatering equipment or heavy pumping in cofferdams, are often covered in the preliminaries. To include the costs in the excavation rate would be distorting.
- Any other plant item we have omitted.

Sundry items — small tools (materials)

The small items which we buy at the local builder's merchant — hammers, trowels, axe handles, nails, etc.—tend not to be included in the rates. Allowance can be either as a percentage of the labour costs on the contract, i.e.

5% of total labour costs

or we can have a weekly allowance which depends on the size of the job. I would suggest:

Small job allowance	£50.00 per week
Medium job allowance	£75.00 per week
Large job allowance	£100.00 per week

Setting-out equipment (plant)

We are all aware of the level, staff and theodolite. In addition we should consider.

- timber for profiles, pins, paint on road contracts
- electronic setting-out equipment
- measuring tapes.

The requirements will clearly depend on the nature of the job itself. Typical cost allowances are.

Level, staff and theodolite	£35.00 per week
Electronic instruments	£100.00 per week each
Tapes	£100.00–£150.00 per contract.

Signs (materials)

- A major sign, say 3 m × 3 m in size, can cost £1500.00 to provide and erect, and two will be required.
- Company signs, say 2.4 m × 1.2 m, can cost £300.00 to provide and erect (2 uses).
- Smaller signs, say 1.2 m × 0.6 m can cost £100.00 to supply and erect (2 uses).
- Price smaller signs pro rata. A small fingerboard is likely to cost £15.00 to supply and erect.

Temporary works (labour, plant, materials)

Provision for items such as:

- shoring in excavations
- temporary bridging
- temporary cofferdams

are priced where necessary as specific items for each contract.

Temporary roads (labour, plant, materials)

There may be a requirement for site access roads in areas where the existing ground cannot take the pressures exerted by site traffic. To strip topsoil, provide a 150 mm thick hardcore standing on a Terram textile base and to reinstate at the end of the job will cost about £8 per m^2.

Protection of the existing location (labour, plant, materials)

- Existing properties may require temporary shoring when a new basement is constructed adjacent to them.
- Gas mains etc. may require support or a protective concrete slab to cover them.
- Existing roads may require regular sweeping. A brushing and cleaning lorry costs around £35 per hour and the minimum hire period is 4 hours. You need to look at the site, to assess the measures necessary, and then to price them.

Testing (sub-contract)

This usually applies to concrete test cubes which are produced on site and taken for testing at a local approved laboratory.

Allow £6 per cube

Beware when there is a requirement for a testing laboratory on site, especially if soils testing is involved as the frequency of such tests often requires the presence of an engineer and a technician full time on site. The equipment itself is also expensive.

On a £1 million earth dam the testing costs were

£1500 per week for staff and cars
£2500 for hire of company owned testing equipment.

Security (sub-contract)

The present day practice is to use security firms either to make regular visits, or to keep a permanent presence on site. A quote will be provided based on individual requirements. A charge of £5.50 per visit, with three visits per night is reasonable. This covers a patrolman visiting the site and walking it at various intervals. Each visit takes around 30 minutes.

Winter working (labour and plant)

The effect of weather on a construction site will vary, dependent on location. The reduction of working hours in winter will affect all sites. A site in the fields is likely to be affected more than one on hard-standing in a built-up area. River work is often simple in summer, but virtually impossible in winter.

Workpeople become less efficient when they are cold, wet and are wearing protective clothing.

Two examples follow: one covering work in fields, the other is for a built-up area on a permanent hardstanding. The figures used, and the time loss calculated are likely to be the *minimum* time loss compared with summer working. Work in fields or on rivers can, on occasion, be impossible, while work in the northern areas of England and in Scotland can be affected to a greater extent than is shown here.

Winter work in fields. The following assumptions are made initially:

- working hours cut from 50 to 40 per week from mid-November to mid-February
- 10% output loss due to cold, wet weather and wearing protective clothing
- effect commences 1 October and ends 31 March.

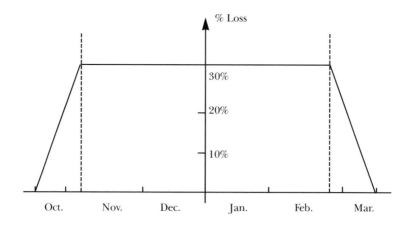

Fig. 2. Winter work in fields

Time lost is

> mid-November to mid-February
> 13 weeks × 0.3 = 3.9 weeks

> 1 October to mid-November and mid-February to 31 March
> 13 weeks × 0.3 × $^1/_2$ = 1.95 weeks

> Total time lost = 5.85 weeks.

Winter work in roads or on hardstandings. The following assumptions are made:

- working hours cut from 50 to 40 per week from mid-November to mid-February
- 10% output loss due to cold, wet weather and wearing protective clothing
- effect commences 1 November, ends 28 February

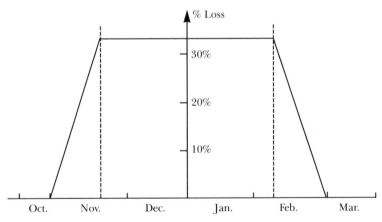

Fig. 3. Winter work in built-up areas

Time lost is

> mid-November to mid-February
> 13 weeks × 0.3 = 3.9 weeks

> 1 November to mid-November and mid-February to end February
> 2 × 2 weeks × $^1/_2$ × 0.3 = 0.6 weeks

> Total time lost = 4.5 weeks.

While these results are empirical, they do show how to carry out your own calculations for loss of time due to winter working.

Temporary covers (sub-contract)

This covers contracts where structures may need to be 'cocooned' for the winter or other period. Scaffold will already be covered separately in the preliminaries. Here we can allow the following:

- the cost of sheeting the scaffold, which will be a sub-contract quotation
- a cover may be required over the structure, supported by scaffolding. This will also be a sub-contract quotation.

A further type of temporary covering involves the use of temporary mobile covers to protect newly completed work — concrete on airfield runways, for example.

Site transport (plant)

Most sites utilize one or more of the following items.

- A dumper — charges vary dependent on size, but typical examples are
    $1^1/_2$ t at £70 per week
    3 t at £100 per week
- A wheeled loader — for moving bricks and other heavy items and lifting them onto scaffold
    weekly charge — £250 for a 2 t four-wheel drive machine
- a road contract may utilize Land Rovers or similar vehicles
    weekly charge — £220 per vehicle.

Clear site — (plant and labour)

- A loaded muck skip, provided and removed from site when full costs around £100.
- The site may require regular clearing, chargeable at existing labour rates.
- It is always necessary to have a cleaning gang go over the site at the end of a job to give things a final clean. This would be chargeable at existing labour rates.
- It is usual to carry out snagging of completed work at the end of the maintenance period. This will also be chargeable at existing rates.

Insurance (materials)

Insurances vary from one company to another and between building and civil engineering work. Typical costs are:

- building insurance   0.3% of contract value
- civil engineering insurance   0.8% of contract value
- marine insurance   3.5% of contract value upwards
- floating craft insurance   4% of floating plant value upwards.

Bond (materials)

The amount of Bond insurance payable is dependent on the financial status of the organization. Clause 10 of the *ICE Conditions of Contract 6th Edition* envisages a Bond not exceeding 10% of the contract value.
 While the cost varies, a sensible charge would be

£1 per 1% of 10% of the contract value per annum

On a £1 000 000 contract with a two year contract period and a one-year maintenance period, the Bond cost on this basis would be

£1 × 1/100 × 1/10 × £1 000 000 × 3 years
cost × 1% of 10% of contract value × number of years
= £3000.

Inflation (labour, plant, materials, sub-contractors)

We need to examine the costs of inflation against the following background:

- labour costs increase on an annual basis of, say, 5% at end of June
- plant costs increase very little on an annual basis
- sub-contract prices include their own allowance for inflation, so no further allowance needs to be made
- permanent works materials can be purchased early in many instances.

 Take an example of a contract which consists of

£200 000 labour
£300 000 plant
£500 000 materials
£1 000 000 sub-contractors

This allows for the elements of these resources which are included in the preliminaries.

Job starts on 1 January, on a 52-week programme.
Assess inflation.

Labour
Increase halfway through contract (end June) of 5%. Say this increase affects 50% of total labour.
Therefore cost = $^1/_2 \times £200\,000 \times 5\% = £5000$

Plant
Prices assumed static this year.
Therefore cost = Nil

Materials
Purchase reinforcement, pipes at start of job
£100 000 fixed cost + national agreed prices on £200 000
Therefore inflation on remaining £200 000 at 5% p.a. average purchase six months ahead
Cost = $£200\,000 \times 5\% \times {}^1/_2 = £5\,000$

Sub-contractors deemed fixed as main contract. Road surfacing is a problem.
Therefore allow £50 000 surfacing at 2% p.a. £2000.

On this basis, the inflation allowance on a £2 000 000 contract is £12 000 on 0.6%.

Summary sheet

The items we have just considered are listed on a summary sheet and totalled in their respective resource elements, as shown in Table 12.

For the site set-up costs, or preliminaries, we now have the totals of the required costs for:

- labour
- plant
- material
- sub-contractors.

These totals will now be added into the contract total.

Table 12. Summary of site set-up costs or preliminaries

Item	Labour	Plant	Materials	Sub/ contractors	Total
Staff					
Attendant labour					
Cars and vans					
Offices, canteens, stores and toilets					
Services					
Compound					
Health and safety					
Sub-contract attendance					
Travel time					
Subsistence					
Plant					
Small tools					
Setting out costs					
Signs					
Temporary works					
Temporary roads					
Protection					
Testing					
Security					
Winter working					
Temporary coverings					
Site transport					
Clear site					
Insurance					
Bond					
Inflation					
Fees					
Totals					

4

Calculating rates

In Chapter 1 we considered:

- the different types of estimating and the differences between them
- the need for practical experience of a particular type of work before starting to price it
- the importance of pricing civil engineering work using first principles
- how to start pricing large contracts
- how to avoid some of the errors which people make in their approach to the estimating process
- how items are grouped for pricing purposes.

The intention was to indicate how to progress the estimating process and the pitfalls to avoid. Chapter 2 introduced us to the elements of the estimating pricing system. We considered the resources of:

- labour
- plant
- materials
- sub-contractors.

We also gained an insight into the pricing of:

- labour
- plant
- materials

Our future prices will be built up using the resources of Labour, Plant, Materials and Sub-Contractors and we will apply the rates used in this chapter.

In a Live Tender Situation all rates will be different to those used here. However, the principles behind what we have used and how we use them will always remain the same. And it is the principles which concern us.

Chapter 3 gave an insight into the pricing of the requirements for the site set-up itself—the preliminaries.

In Chapter 4 we will see how to price work using these resources. While we will use the basic rates which were set out earlier for the elements of Labour, Plant and Materials, the 'live' tender situation will use:

- labour—the standard rates of the business
- plant—the standard rates of the business and the local market-place
- materials—prices from local suppliers. We will also review the amount of wastage and other costs as we examine different types of work.

USING THE RIGHT LAYOUT TO PREPARE OUR ESTIMATE

I have looked over people's shoulders at the estimates they were preparing for a long time and have seen some dreadful mistakes made quite unnecessarily. Mistakes which could have been avoided if people had observed a few simple disciplines.

- Ensure that everyone in the team uses the same system and understands it. I have seen total confusion occur when some people used the labour column for materials prices, for example.
- You must work neatly so that you can think neatly and for others to understand and check that what you are doing is correct. I have always seen untidy work lead to confusion and inaccurate estimating.
  - Use pencil, not ball-point pen. Errors can be rubbed out and neatness is maintained.
  - The sheets of paper you use to work up a price need to be lined. Unlined paper always leads to untidy work.
- It is important, both in the tender preparation and future contract use, that the element prices for the Labour, Plant, Materials and other resources are not confused with each other. At tender stage they need to be kept clearly separate.

Fig. 4. Sample estimating sheet

White Rose	Estimating Sheet			Estimator: EMH	
	Project: Project No.:			Date: 1–1–02 Sheet No. 1	
Item	Description	L	P	M	S/C
1 20 mm	Concrete C.15 Supply etc	20 mm	20 mm	20 mm	20 mm

Figure 4 gives a suitable estimating sheet layout. I would recommend each estimating organization to develop its own estimating form. Forms should be preprinted to ensure that everyone adopts the same technique. An A4 format is realistic. To state the obvious, the letters L, P, M, S/C stand for the four resources mentioned several times previously. Their use is standard practice. The 20 mm wide columns work quite well.

I remember checking the first tender which a new estimator in our unit had prepared. His figures in the L (labour) build-up were quite high. It transpired that this was his materials column and my conclusion had to be rapidly revised. It reinforced the point of having a standard system, in this instance a method of presentation, which we all use.

CALCULATING RATES

No price can be sensibly prepared without:

- a visit to the site of the proposed works
- a properly measured bill of quantities
- a knowledge of the specification for the work being priced
- an assurance that the conditions of contract are standard.

These prerequisites are considered individually below:

The site visit

The drawings will not tell you everything about the site which you need to know for construction purposes. Examples of aspects of the site which will not be covered by drawings follow.

- A stream. How wide is it, how fast is the flow? Will a pipe crossing or a temporary bridge be needed?

- Ground conditions. Is the ground firm? Can wheeled vehicles be driven on it? Or are tracks needed? Are temporary access roads required?
- Where is the nearest tip? This will affect the excavation price.
- Adjacent properties. Are they in a good state of repair? If they need support, will it be a simple matter? Or will it require painstaking effort?
- Traffic conditions. Heavy traffic on adjacent roads can have a significant effect. If you are doing a road job it can be a critical issue.

The site visit is crucial to help you decide how best to carry out the work and the actions you need to take to ensure that you will be able to carry it out. It is good practice to have a construction colleague and perhaps the plant manager on the visit with you. Their opinions of what can be done and how it should be done in the particular circumstances will help you to set about pricing any work with confidence.

The bill of quantities

A Bill of Quantities is essential to any pricing exercise. The more accurate the Bill, the more accurate the price. If a Bill of Quantities is not provided, then you need to produce your own.

You cannot price the Material requirement without a bill of quantities. Nor can you get an accurate price from Sub-contractors (unless you make them responsible for providing their own Bill of Quantities and pricing that). Your Labour and Plant prices will also be approximate. An approximate Bill will only give you approximate values of the work required.

A complete and accurate Bill of Quantities provides the Estimator with the best opportunity of calculating a complete and accurate price.

The drawings

Only a full set of Drawings will give you full details of what is required. While the Bill of Quantities will tell you how much of everything is required, it will not tell you how complex the work is. Only the Drawings will show whether:

- reinforcement is easy to fix and repetitive, or is complicated
- the concrete sections are standard, or all different sizes.

The drawings are essential to assess the degree of buildability of the work.

The specification

The specification defines the requirements of the Engineer on site. These have to be priced. The specification also defines the required standards for the contract. It affects all Material prices as it specifies the quality standards required. It affects Sub-contract prices by setting the standards to be worked to. The required standards also affect programme considerations and the costs of the labour and plant resources required on the site.

To summarize, before you start to calculate any rate you should have

- full information on the contract requirements
- a knowledge of the actions required to control physical conditions on the site itself
- the basic prices of the resource requirements of the job.

Once you have this data, your practical experience will give you an appreciation of the resources needed to carry out the various parts of the work and for how long you will need each resource. The prices are then calculated as:

- materials—cost per unit (m^3, m^2, m, no., etc.) × number of units
- labour—no. of man-hours × cost per man-hour
- plant—no. of machine hours/days × cost per machine per hour/per day
- sub-contractors price the elements of Labour, Plant and Materials in exactly the same way as you are going to do, or as the Estimator now does. You use the relevant rates provided by the various Sub-contractors.

Let's now look at some examples of how Net Prices are built up using the prices we obtained in Chapters 2 and 3. The intention at this stage is that you fully understand the net pricing system.

Example A A simple concrete structure consists of concrete, reinforcement and formwork. The quantities and prices are as follows.

Quantities
Concrete 600 m^3
20 mm reinforcement 40 tonnes
Formwork 300 m^2

Prices

C30 concrete	£61.67 per m^3
C15 concrete	£68.48 per m^3
20 mm MS reinforcement	£287.00 per t
20 mm plywood	£6.10 per m^2
Timber, planed all round (PAR)	£395 per m^3
Wall ties	£1.00 each
Bolts	£1.50 each
Cones	£1.00 each
Fixing bolts and plates	£2.50 per set
Labourer	£8.00 per hour
Tradesman	£9.25 per hour
30 t crawler crane	£865.00 per 39 hours + £150 on/off site
180 cfm compressor	£130.00 per week
Compressor tools	£30.00 per week
Concrete skip	£20.00 per week

Item	Description		L	P	M	S/C
1.	*Supply concrete*					
	C15	£68.48 per m^3				
	extra waste	nil				
	part load charge	£11.00 per m^3				
	standing charge	nil				
	net price	£79.48 m^3			79.48	
	C30	£61.67 m^3				
	extra waste	nil				
	part loads	nil				
	standing charges	nil				
		£61.67 m^3			61.67	
2.	*Place C15 blinding*					
	Assume 4 m^3 per pour and allow 1 hour to place 4 m^3					
a)	lay timber screed 1 carpenter hour +		9.25			
	materials					
	4 m^3 = 40 m^2 × 100 mm—13 m × 3 m					
	32 m of 75 mm × 25 mm = 32 mm @ 75 p					
	m × 1/3 (3 uses)				8.00	
b)	prepare base for concrete—					
	2 men × 2 hours @ £8.00			32.00		
c)	place C15					
	crane + 4 men × 1 hour @ £8(L) × 4		32.00			
	+ crane—1 hour @ 22.00			22.00		
	per 4 m^3 =		73.25	22.00	8.00	
	therefore 1 m^3		18.31	5.50	2.00	

Item	Description	L	P	M	S/C
3.	*Place C30 concrete* Formwork critical to timing (see item 5 build-up) therefore 6 weeks required to place 600 m^3 *Labour* blow out, scabble, pour concrete etc. 3 men × 6 weeks × 45 hours @ £8.00/hour	6480.00			
	Plant crawler crane on /off site—2 @ £150 hire 1/3 × 6 weeks × 45 hours @ £22.00 (2/3 to be charged to reinf. and fwk)		300.00 1980.00		
	compressor on/off site 6 weeks hire @ £160 (incl. tools)		100.00 960.00		
	concrete skip on/off site 6 weeks hire @ £20.00		50.00 120.00		
	Total per 600 m^3 therefore per 1 m^3	6480.00 10.80	3510.00 5.85	– –	– –
4.	*Reinforcement supply and fix* supply 20 mm dia. price per t. (includes waste, packers, tying wire). Sub-contract fixing price per t (from quotation)			287.00	135.00
	cranage for 40 t = 1/3 × 6 weeks × 45 hours × £22.00 40 tonnes = £1980 therefore 1 t =		49.50		
	therefore price per t =	–	49.50	287.00	135.00
5.	Make, fix and strike formwork L formwork—300 m^2 @ 2 man-hours/m^2 = 600 man hours total 2 men × 50 hours × 6 weeks @ £9.25 (includes making)	5550.00			
	P cranage 1/3 × 6 weeks × 45 hours @ £22.00		1980.00		
	M Plywood—3 uses therefore 100 m^2 @ £6.10 backing for 36 sheets ply (2.88 m^2 each) = 36 no. x 18 m per sheet = 648 m @ £1.00			610.00 648.00	
	M soldiers 36 panels × 3 no per panel × 2.5 m long twin 150 × 50 m = 2 × 36 × 3 × 2.5 = 540 m 540 m of 150 × 50 @ £1.50 per m wall ties 150 m^2 wall 1 tie per 2 m^2 = 75 ties @ £1.00 she-bolts—36 panels × 6 @ £1.50 cones—36 panels × 6 @ £1.00			810.00 75.00 324.00 216.00	
	total per 300 m^2 = therefore 1 m^2 =	5550.00 18.50	1980.00 6.60	2683.00 8.94	– –

The important points to learn from this example are

- neatness and explanation of the process
- taking a step at a time. If you can also understand the arithmetic, that is a bonus.

Example B   I was providing a course on estimating to the engineers of a mining company. They had some experience, knew the prices and outputs of the equipment which their business used so we decided to price a major open-cast coal operation in one hour.

The points to note are:

- large quantities make tidy estimating vital
- large quantities should not be off-putting, you just put larger numbers into your calculator.

The sequence of work is listed (items 1 to 10). The sequence needs to be clearly stated and each sequence item noted. You then price as items 1, 2, 3, 4 etc. until each sequence item has been covered. The example we took was as follows.

- 1 km × 1 km site
- excavate 10 m to seam 1 which is 1 m thick
- excavate 10 m to seam 2 which is 1 m thick

Recover coal over area 800 m × 800 m, working 100 hours per week.

Sequence of work

(1) Strip topsoil 150 mm thick — $800 \times 800 \times 0.15 = 96\,000$ m^3
Scraper on/off — £1000 each way
Scraper strips 100 m^3/h at hire cost of £125/per hour

(2) Subsoil strip — 2 m deep × $800 \times 800 = 1\,280\,000$ m^3
Scrapers — 150 m^3/h at cost of £125/per hour
10 scrapers required

(3) Drill to first coal seam
8 m deep holes at 4 m c/c = 200 × 200 holes
40 000 holes, 8 m deep — drill 1 hole per rig hour
Rig costs £30 per hour + 3 men at £9 per hour each
Gelignite use 4 oz per m^3 at £5 per 1 lb wt

(4) Excavate to first seam after blasting
(D/L) dragline on/off site and set up — quote £30 000
Excavate first cut 30 m wide, load to dump trucks and cart to far side of site
Volume 800 × 30 × 8 m, i.e. 10 m to seam 1 less 2 m sub-soil strip

D/L — £300 per hour — excavates and loads at 200 m³/ per hour
Truck — £80 per hour
2 km haul and return – $^1/_2$ hour trips carrying 20 m³/trip in
each of 5 trucks, i.e. 200 m³ per hour

(5) Excavate seam 1 coal and cart to washery
(54 RB) large digger on/off site — £5000 (quote)
Vol = 800 × 800 × 1 = 640 000 m³
Tonnage 640 000 × 1.8 t/m³
54 RB loads at 200 m³/hour = 360 t
1 hour round trip carrying 20 t per lorry
therefore, say 18 lorries + breakdowns — 20 no. lorries
54 RB at £100/per hour
Lorry at £25 per hour

(6) Excavate remaining cuts and fill into earlier cuts using D/L only
800 × 770 × 8 m
D/L excavates and dumps 300 m³ per hour
Costs £300 per hour

(7) Drill to second coal seam
200 × 200 holes × 10 m deep
Working rates and costs as before

(8) Excavate first cut and load into trucks and cart to rear of site
As before for first level (item 4) + 20% for depth (bringing trucks
out on ramps, making ramps, etc.)

(9) Excavate rest of second level dig and cast into previous cuts — as
(6)

(10) Excavate second coal seam
As before + 20% for depth
Add (D10) large bulldozer to assist lorries on ramps
D10 — £2000 on/off site + £100 per hour hire

Item	Description	L	P	M
1.	Strip topsoil 800 × 800 × 0.15 = 96 000 m³ 10 scrapers on/off @ £2000 each strip @ 100 m³ per hour — 100 hour week 10 scrapers × 100 hours @ £125 1 WEEK		20 000 125 000	
2.	Strip Subsoil 800 × 800 × 2 = 1 280 000 m³ scraper hours @ 150 m³ per hour = 8533 hours @ £125 10 scrapers for 8$^1/_2$ weeks 8$^1/_2$ WEEKS		1 066 625	
	C/F		1 211 625	

53

Item	Description	L	P	M
	B/F		1 211 625	
2a.	Banksmen 10 weeks × 2 shifts × 50 hours @ £9 × 6 no.	54 000		
3.	Drill 1st seam after soil stripped 40 000 holes @ 1 hole per hour based on 100 hours week 400 drill weeks 8 no. drills × 50 weeks 8 no. drills on/off @ £1000 40 000 hours @ 27.00(L) + £30.00(P) gelignite—800 × 800 × 8 = 5 120 000 m³ 5 120 000 × $^1/_4$ lbs @ £5 per lb = 50 WEEKS	1 080 000	8000 1 200 000	6 400 000
4.	Excavate first cut and load & cart D/L on/off and set up (quote) volume = 800 × 30 × 8 = 192 000 m³ excavate @ 200 m³/hour = 960 hours — 10 weeks × 100 hours Dump trucks 5 no. on/off D/L + 5 trucks × 10 weeks × 100 hours @ £700 (£300 + 5 @ £80) 4 banksmen × 2 shifts × 10 weeks × 50 hours @ £9 D10 level spoil at dump point (price as item 10) on/off site 10 weeks × 100 hours @ £100/hour 10 WEEKS	36 000	30 000 5000 700 000 2000 100 000	
5.	Excavate seam 1 coal and cart to washery etc. 54 RB on/off site excavate @ 200 m³/hour volume = 640 000 m³ —3200 hours = 32 weeks 3200 hours × 54 RB @ £100 hour 20 LORRIES @ 500 hour 3200 hours @ £600/hour Banksmen 4 no. × 2 shift × 32 weeks × 50 hours × £9	115 200	5000 1 920 000	
6.	Excavate by dragline overburden to seam 1 of coal volume = 800 × 700 × 8 m deep output is 300 m³ per hour dragline costs £300 per hour cost = $\dfrac{800 \times 700 \times 8}{300}$ × £300 ADD banksman × £9 per hour $\dfrac{800 \times 700 \times 8}{300}$ × £9	134 400	4 480 000	
	C/F	1 419 600	9 661 625	6 400 000

54

Item	Description	L	P	M
	B/F	1 419 600	9 661 625	6 400 000
7.	Drill rock overlying 2nd seam, charge and blast 200 no. × 200 no. holes × 10 m deep, use labour and plant as item 3 which we priced previously Drill to 2nd seam 40 000 holes × 10 m deep, allow $1\frac{1}{2}$ hours per hole per rig based on 100 working hours per week and 8 drill rigs. Time taken = 60 000 drill hours			
	60 000 @ £27.00 (L) and £30.00 (P) Gelignite 800 × 800 × 10 = 6 400 000 m^3 6 400 000 m^3 × $\frac{1}{4}$ lb @ £5 per lb	1 620 000	1 800 000	8 000 000
8.	Dig first cut of rock, load and cart to far side of site volume = 800 × 30 × 10 m excavate @ 200 m^3 per hour truck does 2 trips @ 20 m^3 per hour therefore 5 trucks excavating time — $\dfrac{800 \times 30 \times 10}{200}$			
	= 1200 hours @ £300.00		360 000	
	5 no. dump trucks × 1200 hours @ £80		480 000	
	4 banksmen × 1200 hours @ £9	43 200		
	D10 levelling spoil 1200 hours @ £100		120 000	
9.	Excavate remainder of 2nd level. Dig and cast aside (as 6) volume = 800 × 700 × 10 m deep @ 300 m^3 per hour, D/L costs £300/hour cost = $\dfrac{800 \times 700 \times 10}{300}$ @ £300		5 600 000	
	ADD banksman @ £9 per hour 18 666 hours @ £9—say	168 000		
10.	Excavate 2nd seam of coal (refer to note — price 5 + 20%) + D10 to assist lorries			
	Price 5 was	115 200	1 925 000	
	ADD 20%	23 040	385 000	
	ADD D10 3200 hours (from 5) + 20% = 3840 hours @ £100		384 000	
	NET TOTAL ($\sum$ 1 to 10)	3 389 040	20 715 625	14 400 000

Even if you have never visited an open cast coal site before, if you were to be escorted around a working site by an experienced colleague who knew methods, outputs and prices, and gave you a full brief on how the job worked, I suspect it would be perfectly feasible for you to then price the above exercise.

When you are working with large quantities and sums of money, it is easy to make errors. You may put a decimal point in the wrong place or the size of the numbers may confuse you. To simplify matters, work to the nearest whole unit. For example:

- quantities—work to the nearest whole unit
- sums of money—32 674 × £99.12 = £3 238 646.88, express as £3 238 647
- it is standard practice to charge Plant rates to the nearest pound.

As an indication, check the arithmetic used in the following example.

Example C A twelve storey block of flats is to be refurbished. Refurbishment involves new kitchens, bathrooms, double glazing and redecoration. Access to the work is via an external scaffold with a hoist for carting materials up and down. A 'fetching and carrying' gang is available to bring new materials to the work point and remove dismantled items.

Attempt to price the double glazing provision.

Information available is as follows.

- The scaffold, hoist and the fetching and carrying gang are priced elsewhere.
- There are 200 windows, each 1.5 m × 1.5 m, to be replaced.
- The existing windows are single glazed, metal framed.
- A new unglazed frame, 1.5 m × 1.5 m, costs £200.
- There are four double glazed pieces, each 0.75 m × 0.75 m, to each large frame. These cost £25 each.
- Sealant between the new frame and the structural surround costs £10 per window.
- Tradesmen cost £9.25 per man-hour.
- Van and tools cost £8.00 per hour.
- Operatives cost £8.00 per man-hour.

As a useful starting point the operations involved and the questions to be asked are listed below.

(a) Imagine a window 1.5 m × 1.5 m. You are going to take it out using the scaffold as a platform. How many men do you need, and for how long?

(b) Now put in the new 1.5 m × 1.5 m frame. How many men do you need and for how long do you need them? Fixing is via bolts drilled through the frame and into the surround and the frame is to be plumb.

(c) The new frame is a single unit divided into four quadrants and a double glazed unit fits into each quadrant. How many men and for how long do you need them to complete one window?

(d) The sealant used to seal the new frame to the structural surround is an expanding foam material, squeezed into the gap. How many men are needed and for how long to complete one window?

Do not forget that there are 200 windows. A possible answer is given at the end of the chapter.

Example D   In Chapter 3, the costing of the site set-up or preliminaries was covered. Now consider the details of the preliminary items provided on a job with which you are familiar, or build up an imaginary provision.

An extension of time of 13 weeks has been granted. What is the net cost of the preliminaries on your site for the 13-week period?

Ask a colleague to check your answer and make sure that you work neatly.

Grouping items for pricing.   In Chapter 1, the method of 'grouping' items for pricing was explained, where similar items are considered globally and a global cost worked out. This global cost is then fed back into rates, and some rates will be higher than others.

Example E   In Example A, a net rate for *placing* class C30 concrete was calculated. This was an *average* rate. If we imagine a bill of quantities for the concrete, which is based on *CESMM 3*, it could look like the following.

Item	Description	L	P	M	S/C
	Placing C30 concrete AV PRICE per m^3	10.80	5.85	–	–
F624	*In ground slabs*, footings etc. exceeding 500 mm thick (this will be quite cheap, reduce average price by 30%) per m^3	7.56	4.10	–	–
F633	*In suspended slabs* 300–500 mm thick assume average price per m^3	10.80	5.85		
F644	*In walls* Exceeding 500 mm thick assume average price per m^3	10.80	5.85		
F654	*In columns*, cross-sectional area 0.25–1 m^2 (This will be expensive, increase rate by a factor of 2) per m^3	21.60	11.70		

Item	Description	L	P	M	S/C
	Now consider the reinforcement. We worked out a price to supply and fix 20 mm dia. mild steel. Sub-contract prices are obtained for 16–32 mm dia. We will allow the same labour and plant costs for each diameter.				
G516	20 mm MS reinforcement per tonne	–	49.50	287.00	135.00
G517	25 mm MS reinforcement per tonne	–	49.50	287.00	130.00
G518	32 mm MS reinforcement per tonne	–	49.50	300.00	125.00
	In terms of the formwork, we calculated the rate per m^2 based on formwork being vertical.				
G244	That is: vertical formwork 0.4–1.22 m high per m^2	18.50	6.60	8.94	
	We may also have to consider formwork such as:				
G235	battered exceeding 1.22 m high (add 20% to vertical rate) per m^2	22.20	7.92	10.73	
G281 to G286	Components of constant cross-section— beams, columns, etc. (add 10% to vertical rate), per m^2	20.35	7.26	9.83	

Example F We may have a contract where we have to put pipes of many diameters through concrete walls. On a water treatment plant there could be over 100 such pipes. We *don't* price every pipe.

Description

Take as an example:

pipe diameters   75 mm–10 no.
100 mm–15 no.
150 mm–30 no.
200 mm–6 no.
300 mm–8 no.
400 mm–20 no.
500 mm–10 no.
600 mm–10 no.
750 mm–10 no.
900 mm–8 no.
1200 mm–5 no.
1500 mm–3 no.

We want a simplified net pricing structure.

Item	Description	L	P	M	S/C
	Say, pipe dias. 75–200 mm are similar. average 150 mm. dia. We have to:				–
	– box out in wall for pipe prior to concreting. Use 'expamet', box out 450 × 450 mm				
	1 man × 1 hour + 'expamet'	9.25		3.00	
	– set pipe in position 1 man × $\frac{1}{2}$ hour	4.62			
	– put fwk over void				
	1 man × 1 hour + formwork (1 m^2)	9.25		9.00	
	– place concrete (0.1 m^3)				
	1 labourer × 1 hour	8.00			
	For 150 mm dia.	31.12		12.00	
	Say, pipe dias. 300–600 mm dia. are similar, average 400 mm dia. We have to:				
	– box out in wall 1 m × 1 m				
	1 man × 2 hours + expamet'	18.50		6.00	
	– set pipe in position				
	2 men × 1 hour + lifting tackle	18.50		2.00	
	– formwork over void				
	1 man × 1 hour + formwork (2 m^2)	9.25		18.00	
	– place concrete (0.4 m^3)				
	2 men + crane × $\frac{1}{2}$ hour	8.00	11.00		
	For 400 dia.	54.25	11.00	26.00	
	Say, pipe diameters 750–1500 mm dia. are similar, average 1200 mm dia. say:				
	– box out 1.75 × 1.75 m				
	1 man × 2 hours + expamet	18.50		10.00	
	– set pipe in position				
	2 men + crane × $1\frac{1}{2}$ hours	27.75	33.00		
	– formwork to void (4 m^2)				
	1 man × 2 hours + formwork	18.50		36.00	
	– place concrete (say 1 m^3) (3 men + crane × $\frac{1}{2}$ hour	12.00	11.00		
	For 1200 mm dia.	76.75	44.00	46.00	
	Now look at the various diameters				
	75 mm	15.00	–	8.00	
	100 mm	23.00	–	10.00	
	PRICED 150 mm	31.12	–	12.00	
	200 mm	39.00	–	17.00	
	300 mm	47.00	–	22.00	
	PRICED 400 mm	54.25	11.00	26.00	
	500 mm	58.00	17.00	30.00	
	600 mm	63.00	23.00	34.00	
	750 mm	68.00	29.00	38.00	
	900 mm	72.00	36.00	42.00	
	PRICED 1200 mm	76.75	44.00	46.00	
	1500 mm	80.00	52.00	50.00	

*Rates obtained by interpolation

Net rates and examples will be covered later in this book. The intention in this chapter has not been so much to *calculate* rates themselves, but to demonstrate how they are calculated and how you can simplify the pricing of various groups of rates. The method I have shown, while perhaps being simple and certainly obvious, is fundamental to any pricing exercise.

The prices which you use will be those common to your organization. If you are inexperienced in certain areas, you could use internal output records. Whenever you use data provided by others however, you must ensure that the data is *relevant* to the case you are pricing.

Different pricing methods will be considered in Chapter 5. But first, a possible answer for Example C, which we looked at earlier in this chapter, is given below.

Item	Description	L	P	M	S/C
	Example C				
(a)	Remove 200 existing frames. 1 labourer + 1 tradesman take 1 hour per window Overall cost = 200 × £17.25 (L)	3450			
(b)	Put in 200 new frames, including supply. Allow 1 labourer + 1 tradesman @ 1 hour per frame = 200 × £17.25. Supply new frames 200 × £200 (M)	3450		40 000	
(c)	Fix new double glazed units. Supply 200 windows × 4 no. per window × £25 Fix 200 no. × 4 no. @ $^1/_4$ hours each × £9.25 (tradesman)	1850		20 000	
(d)	Seal new frames to structure. Sealant 200 no. × £10 Tradesman 200 @ $^3/_4$ hours × £9.25	1388		2000	
(e)	Clean new window and frame Labourer 200 no. @ 1 hour × £8.00	1600			
(f)	Van cost—time required using a 5 man gang is 2 hours per window (say) (work parts overlap each other) Van cost = 200 no. × 2 hours × £8.00		3200		
	Therefore net total	11 738	3200	62 000	–

5

The sources of our prices

Any price which we consider has been calculated on the basis of the cost of resources used and an added element for overhead and profit.

Consider the scenario of goods in a shop. There is little opportunity for problems to arise. The purchaser sees the goods and can examine them. The price displayed is current at that time. The person selling (the vendor) can change the price at will. At the time of purchase there is, theoretically at least, little reason for either party to be dissatisfied. The buyer wants the product and is happy about price, the seller is happy to trade the goods in return for that price.

A quotation for the supply of a service or a product, prepared to suit set requirements, is slightly different. The purchaser cannot see the final product before purchasing. The vendor may have got the price wrong or misunderstood the requirement. In most cases, where quotations are for specific products made to recognized standards, there are few problems. The purchaser knows the product, the vendor has the price right due to the routine nature of the request. This nearly parallels the goods in the shop situation. Where quotations are for a non-standard item, defined by the requirement of the purchaser, then we are more likely to encounter problems. If the person preparing the quote discusses the requirement carefully with the proposed purchaser, then errors are less likely. By the same token, the purchaser can try to ensure that those quoting know exactly what they have to allow for.

A schedule of rates, or the list of charges for different types of supply, is rather different. The list of items has no quantities and is not specific to the particular requirement. It works well with standard products — nuts and bolts, sweets, newspapers, products available in grocers or ironmongers. In the context of construction work it can prove suitable, or it can be a disaster. Schedules need careful consideration before they are used.

Books are written by individuals who can only express their own opinions on what they have learnt from their experience. What you learn from the book is only of benefit if the experience of the author is relevant to your needs. As we are looking at the estimating process it is important to understand the following.

- We are seeking the cheapest *final* price. This is the one which decides our profitability. This is not necessarily the lowest initial price we receive.
- Each price is calculated by a person picturing the task, resourcing it and then deriving the cost. You must ensure that the person
  - sees the right picture
  - competently prepares the price
  - and that the price is fully relevant to your needs.

Some of the sources of prices which we use are now examined in more detail.

SPECIFIC QUOTATIONS

Receiving quotations

We receive quotations from materials suppliers, sub-contractors and occasionally a plant supplier. It is best to try to get three or more quotations for each requirement. While you will always favour one of the quotes, the others will hopefully confirm it.

Always check the arithmetic on quotations. This applies primarily to sub-contractors. You need to ensure that the arithmetic is correct. You must also ensure that all items are priced in a quotation which you intend to use.

With materials quotations, you need to be aware of any precautions which have to be taken to comply with the *Care of Substances Hazardous to Health (COSHH) Regulations*. There may be a cost implication.

Under the *Construction (Design and Management) Regulations 1994* you have a duty to ensure that anyone you employ is competent to do the work and will deploy adequate resources, in terms of health and safety, to do the work. In commercial terms, you must reflect this by ensuring adequate resources in the right place at the right time. The resources must perform as required for as long as required.

- Materials suppliers must maintain an adequate supply. If you are unsure of this, use a secondary back-up source as well as a main supplier.
- Sub-contractors must provide adequate numbers, with the required competencies when needed.

Providing quotations

When you provide a quotation to a client, you should protect the business by confirming the data you have used in evaluating your quote. This will usually be the information sent to you by the prospective client. Note any assumptions made or inadequacies in the data provided.

Do not make promises you cannot keep. It can be extremely expensive if you fail.

Ensure that the prospective client can pay you and that payment will be within a sensible length of time.

Ensure that what is being asked of you is reasonable, with regard to sensible specification, performance requirements, deliveries, etc.

A specific quotation is intended to be fully relevant to the work requirement, it should be just that.

SCHEDULES OF RATES

Schedules of rates can be used in a variety of ways. The main types and the points to consider when using them are detailed below.

Schedules of average prices

Client teams often have to provide budget prices for proposed new works. The budget price is used to obtain capital approval for the money to be spent. Once approval is given, tenders can be sought in the normal manner. It is common practice for the budget price to be determined by:

- providing a bill of quantities
- examining rates received in tender submissions for similar work
- putting an appropriate rate against each item in the bill
- multiplying through the bill and adding the various totals to give the budget price.

This is a simple method and can provide a speedy answer. Provided the rate of inflation is low and the economy stable the answer should be reasonably accurate.

If you have high inflation, however, and perhaps a changing market-place as well, then the answer could be very inaccurate. I recall a client telling me that prices had fallen by 40% at the start of a particular recession. This arose from the following factors:

- a 10% contractors' margin had disappeared
- materials inflation at $1\frac{1}{2}\%$ per month had disappeared
- a reduced market forced contractors to discount all materials and sub-contract quotations (the quoted prices would be reduced by the contractor)
- contractors were submitting tenders using shorter programme times.

No schedule of prices would predict this. In addition, any schedule of prices is based on historic data. Without knowing how the schedule was originally built up and any changes which had occurred between the original and the current time, accurate prediction is difficult, if not impossible.

Schedules for pricing

Some clients use schedules of work for contractors to price when they are seeking tenders for maintenance work or small contracts. The client could be a local authority, an industrial undertaking, or a retail business—all regularly use schedules for pricing purposes.

The normal practice is for the client to provide a schedule of work items and ask tenderers to price the schedule. No quantities are given but there is often a statement of what the amount of work in a given year is likely to be.

The problem for the estimator is that there is no indication of how much work is likely to be required in any given visit. An assumption has to be made and common sense dictates that we should assume a small amount at any given time. The effect of assuming a small amount of work is that the calculated rates will be high. The competitive nature of the business usually gives work to the lowest price and this usually means an over-optimistic assessment is required to ensure success.

I found such work virtually impossible to win on a sensibly priced basis. This mechanism for seeking tenders is perhaps best suited to the small tradesman who works from a single van with a core of colleagues. Larger contractors, perhaps able to offer a more in-depth service, find it difficult to compete. Some examples are considered below.

Example A.   Retail and industrial clients want an emergency call-out service plus a facility for doing small amounts of maintenance work. With a term contractor working to a priced schedule of rates, a telephone call will ensure the service. From the point of view of the client the procedure is simple and effective.

From the point of view of the contractor preparing a price, assumptions have to be made. What will the work require?

- A single van, possibly a Transit. This is a sensible guess and does not present a problem.
- How many men will be needed? Clearly, this will vary dependent on the work to be done. A logical minimum would be two men.
- Will the work occupy a full day or a full number of days? If it does, then there may be little problem as we pay for labour and plant on a daily basis. Can the team be absorbed immediately back into operation? If not, then a problem of cost will arise.
- If the work takes less than a day, then it could be impossible to keep the team occupied for the rest of that day.

The likelihood is that the prices for such work will always be high. Contractors doing the work will probably lose money, and the service provided must then suffer.

Example B.   Local authorities require maintenance work of a minor nature, often on roads and bridges. It is highly convenient for the client as the work can be ongoing and there is no need to seek tenders constantly.

If the work is continuous, the successful contractor will keep the resources dedicated to the contract fully employed and there may be few problems as costs incurred are balanced by the ongoing value of the work carried out.

If the work is not continuous, and this is a likely scenario, then costs are ongoing but value is not — a problem arises.

The rates for the work vary with the quantity required:

- to replace one kerb, one square metre of block paving, or one square metre of road construction will take almost as long as if five times the amount of work were carried out.

To state the obvious, an accurate price requires an accurate measure of the work to be priced. If we do not have that measure, then we have to make assumptions on what will be required and build up our price based on these assumptions. The assumption is much more likely to be incorrect than it is to be correct.

Work being offered on the basis of a schedule of items which the contractor then prices should be approached with caution.

Using schedules of rates for tender purposes

Chapter 1 states that the appropriate rates for many standard building work tasks can be reasonably predicted. They vary little between different contractors and are relatively standard in any particular location. They are predictable because the resources used are predictable, and standard for the operation. As a result we can use standard schedules of rates for many building work operations without too much anxiety.

Civil engineering work is very different. Rates vary with many factors, such as:

- location
- access
- ground conditions
- tipping facilities
- contamination
- volume of work
- complexity of the job
- degree of buildability
- time of year (winter or summer)
- tidal effects
- river flows
- competition
- degree of risk
- programme requirements.

It is unlikely that any schedule of rates being used would comply fully with all the imposed conditions on any site. There must therefore be a varying degree of risk if we use schedules in any tendering situation. The risk may be slight on many building operations (all internal finishing trades, for example). It would, however, be major on civil works, especially where complex working conditions exist — exposed work in wintry conditions, river works, tidal situations, flood defence work, for example.

You therefore need to be very careful if you do decide to use a schedule of rates. You must

- ensure that the rates you use are fully relevant to the work you are pricing.

- remember that the rates are historic, they were established in the past. You must ask:
  - how old are they?
  - what were the market conditions like at the time the schedules were made up. Are they the same today?
  - how do I update the prices to make them current? A basic inflation allowance is likely to be incorrect.

DATABASES

I feel that a database, with which the estimator is fully conversant, is an excellent guide. The *CESMM 3 Database* identifies all the items likely to be billed under the CESMM measurement process. No estimator is likely to be familiar with pricing *every* bill item in the *CESMM 3*, so it is likely to help everyone to a varying degree.

Specialist data is provided by practitioners in each of the classes. There is comprehensive data on the costs of materials and the build-up of labour and plant gangs for the various construction tasks. This forms useful guidance.

However, it is not realistic to expect each price always to be relevant. Of necessity, the rates are a guide. Commercial pressures are likely to cause materials prices to vary, for example.

The following points will help you to use this, and other databases, more effectively.

Class A — General items

Insurances will be those of your own business.

Plant and labour items will be relevant to the particular contract which you are pricing and the rates you use will be those of your business.

Class B — Ground investigation

Prices will be from specialist sub-contractors and the rates will vary from job to job. Rates provided are for guidance only.

Class C — Geotechnical and other specialist processes

Prices will be from specialist sub-contractors and will vary from job to job. The rates are for guidance only.

Class D — Demolition and site clearance

Prices will vary with location, quantity, degree of difficulty, location of the tip and many other factors.

Class E — Earthworks

The rates assume average ground and weather conditions, together with open and unrestricted access and movement for machines. Each of these will vary from one contract to another. The volume of the excavation has an effect on price as has the percentage of workspace which may be required. On any one contract the accessibility is likely to vary from one section to another.

Tipping and other charges are excluded from the rates.

Class F — In situ concrete

The concrete prices are for bulk supplies based on site batching. This only applies to a few large contracts. The price for small quantities is based on ready-mixed supplies.

In general practice, concrete is supplied ready mixed. The increase in quantity from this source is likely to give a lower materials price.

The wastage factor is put at 5%. This will vary on site according to the factors which prevail there. It is advisable to add further wastage as follows

75 mm blinding	25%
concrete in trenches	15%
concrete in base slabs	5%
concrete in floors, walls	$2\frac{1}{2}\%$
concrete in columns	5%

The concrete gang is likely to remain constant in terms of labour and plant. Placing rates will vary between large and small pours and between simple and complex ones. You need to remain alert and make allowances as necessary. Examples were given in Chapter 4 (Example E).

Class G — Concrete ancillaries

Formwork. The key factor in the materials price is the number of uses for each formwork panel. The labour and plant elements will be defined by the size of the formwork panels, complexity and other factors.

Reinforcement. Fixing tends to be by sub-contract labour and competition is likely to reduce the rates given for the labour element. The plant content will be defined by how much cranage is allowed. (A site crane will usually cover reinforcement, formwork and concreting operations.)

Classes I, J, K, L—Pipes, fittings, manholes, protection

Prices vary enormously and are affected by considerations such as the following.

- Is the work to be carried out in roads or fields?
- Is there an adequate way-leave?
- Is the ground hard or soft?
- Are the trench sides to be supported or battered?
- Is the work repetitive?
- Is there any traffic to contend with?

Classes M and N—Structural and miscellaneous metalwork

All prices will vary according to factors such as the complexity (or simplicity) of the fabrication work, the amount of work in the order, the required delivery date.

Class O—Timber

The price of the material can change quickly due to currency changes and demand.

Much of the work is of a marine nature and the plant provision (often floating craft) can have a very significant impact on price.

Classes P and Q—Piling and ancillaries

Estimators may have to allow for the provision of a piling carpet of hardcore up to 1 m thick for the sub-contract plant to stand on.

An allowance for pumping away any water arising from the pile bores may be required.

An allowance for disposing of excavated material arisings may be required.

Some types of pile, continuous flight augured for example, can be formed with very large mushroom-shaped heads. The heads are expensive and time-consuming to remove.

Class R — Roads and pavings

Prices will vary according to factors such as

- the degree of separation of the work from the traffic
- the quantity of work to be carried out
- ease of access to the site.

Precast concrete items, kerbs, channels, flags and edgings attract fairly standard prices.

Blacktop prices are dependent on the job and its size but are fairly standard to the specific conditions.

Classes S and T — Rail track, tunnels

This is highly specialized work and specific discussion is required with expert suppliers.

Class U — Brickwork, blockwork and masonry

Materials prices can vary considerably during the economic cycle, as can lead times for delivery to site. The labour element is often provided in the form of sub-contract labour. The prices quoted tend to rise steeply at times of industry upturns.

Ensure that you allow adequate access (scaffold). A fork-lift truck is often provided on large sites. Pallets are then lifted directly onto the working scaffold.

Alteration work in live premises needs careful consideration. Cutting work can be particularly expensive.

Masonry of all types attracts quite varied prices. Ensuring adequate resources for your requirements is important. Companies supplying the service tend to be small.

Class V — Painting

Outputs are fairly predicable and correspond closely to a schedule of rates.

Class W — Waterproofing

Rates vary a little with location and quantity but are usually provided on a sub-contract basis. You could use a schedule if necessary without too much concern.

Class X — Miscellaneous work

Gabions.   The price will vary according to accessibility of the site and the location of the gabions themselves. Accessibility can lead to cost implications as the stone has to be brought to the site in the first place. Gabion placing is often best carried out in conjunction with bank excavation. It may be necessary to place the stone into the gabions in situ rather than load them on the bank and hoist them into position. Prices can vary widely.

Fencing, stiles and gates.   Prices tend to be fairly standard. You could use such prices without too much risk. Work is usually carried out on a 'good weather' basis, so prices become more predictable.

Class Y — Pipe renovation and ancillary work

The work is very specialized and quotes are needed for any prices you use. Prices can vary a great deal and it would be very unwise to use a schedule.

Class Z — Simple building work

Carpentry and joinery work is relatively straightforward and predictable but beware of materials prices for any special units. Labour and plant can be in schedule form.

Window, doors and glazing — beware materials prices. Labour and plant can largely be scheduled.

Surface finishes — prices can be scheduled but watch for unusual tiling or other requirements.

Building services — mechanical and electrical work seems to prove to be a regular problem on building work associated with civil engineering. This may be because many civil engineers do not fully understand what is involved. It is advisable always to use a sub-contract quote and even then to check it exhaustively.

Class ZZ — Alterations

Prices vary with quantity, location, prevailing conditions, location of a tip, any restrictions and other similar factors.

Working in a live environment, and this is often the case with alterations, is quite different to having full occupation of an evacuated area.

The rates given are for a job valued in the range £250 000–£500 000 which is quite large.

Rates assume unrestricted access. The nature of alteration work tends to restrict access, that is, all jobs tend to be somewhat restricted. Delivery of materials to an area of a derelict property is likely to be more expensive than moving the same items the same distance on an open site.

To summarize, I feel that a comprehensive database which gives the resource components of the price as well as the price itself is a useful tool for any estimator. You can easily confirm a price by building-up your own price and comparing the two.

If you *do* use a database remember that

- prices are general rather than specific
- price levels will be sensible, and competition is likely to result in lower prices
- wastage factors vary dependent on the location of the work and its type
- gang sizes vary.

As with all prices, those in the database are based on assumed outputs. You must check that they are accurate with reference to the work you are pricing.

DAYWORKS

Most tenders contain an element called 'Dayworks'. This is essentially payment on a Cost Plus basis. Contractors are paid the cost of the labour, plant, materials and any other incidentals which they use at the time and payment is at rates agreed for the resources mentioned. Clients may decide to provide their own schedules.

Equally, there is a Schedule defined by the *ICE Conditions of Contract, 6th Edition*, clause 56(4). This states that where work is carried out on a daywork basis it shall be paid at the rates and prices set out in the Daywork Schedule included in the contract. If no schedule is included, payment will be at the rates and prices and under conditions contained in the 'Schedule of Dayworks carried out incidental to Contract Work issued by the Federation of Civil Engineering Contractors' current at the date of execution of the daywork.

The current edition of the Schedules became effective on 1 July 1998 and the issuer is the Civil Engineering Contractors Association. These Schedules are examined below.

The Schedules — where to use them

Introductory Note 3 states the following.

The Schedules are prepared for use in connection with Dayworks carried out:

(a) incidental to contract work. The contract work is ongoing and the Daywork is part and parcel of it.

(b) where no other rates have been agreed. Other agreed rates could be either existing relevant rates or rates specifically agreed for the Daywork in question.

(c) they are not intended to be applicable for Dayworks carried out after the contract works have been substantially completed. The Daywork rates in the Schedule will usually be inadequate at this stage.

(d) nor are they intended to be applicable to a contract to be carried out wholly on a Dayworks basis. The Schedule rates would be unduly profitable in many cases.

(e) in cases (c) and (d) the circumstances vary so widely that the rates applicable call for special agreement between client and contractor.

Labour. Add to the amount of wages paid to operatives 148%. The 'amount of' wages consists of

- actual wages and bonus paid
- daily travel allowance
- tool allowance
- any other prescribed payments (e.g. inclement weather).

If this is applied to the rate for a Grade 3 Operative, shown in Table 1 of Chapter 2, the following weekly charges for a Grade 3 Operative are derived:

Total taxable pay	264.60
Add inclement weather allowance	13.23
Add travel time say	8.00
Add training and redundancy $7\frac{1}{2}$%	19.85
Net total	£305.68
Add 148%	£452.41
Weekly total	£758.09

This is for 45 working hours, therefore the hourly daywork rate =

£16.85

This is clearly a premium rate compared to our starting net rate of £8.00, even after we add company overhead and profit to the net rate (say $12^{1}/_{2}\%$).

Materials.   To the net cost of materials delivered to site add

- handling costs
- unloading and storage costs
- $12^{1}/_{2}\%$ (includes overhead and profit).

The estimator would have included the net cost of these items in a normal cost calculation. The $12^{1}/_{2}\%$ addition is realistic. There is no premium price for materials when paid on a daywork basis.

Supplementary charges.   Further costs incurred incidental to the daywork are payable:

- transport for the operatives to and from site and on the site itself is charged at Schedule Rates
- any other charges — tipping fees, professional fees, sub-contractor accounts are payable at cost plus $12^{1}/_{2}\%$
- the cost of providing welfare facilities is chargeable at cost plus $12^{1}/_{2}\%$
- additional insurance is chargeable at cost plus $12^{1}/_{2}\%$
- watching and lighting costs are recoverable at Schedule Rates.

When you price dayworks in a bill of quantities prepared under the *CESMM 3* rules (Class A, items 4.1.1 to 4.1.8) a facility is provided to allow a positive or negative adjustment of the four respective types of daywork charge. This enables the estimator to vary the rates or percentage additions used in the Daywork Schedule.

Plant.   A comparison of the net rates for plant given in Table 2 of Chapter 2 with the equivalent daywork rate shows the following.

Plant Item	Hourly net cost (39 hour week)	Daywork cost per hour
40 t crawler crane	£15.56	£65.23
180 cfm compressor	£3.33	£6.24
1.5 t dumper	£2.31	£4.04
2.0 t fork-lift	£6.03	£10.11
JCB 3C	£7.36	£17.03
24 t tipper lorry	£19.23	£27.41

The rates are exclusive of drivers.

The addition of overheads and profit (say $12^{1}/_{2}\%$) to the net cost still shows the daywork charges attracting a substantial premium.

'Guesstimates'

It is one thing to give an off-the-cuff price (a 'guesstimate') for a contract's value. It can, for example, give an estimating team an indication of how much work the contract will generate. The directors need such data to assess how busy the industry, the company or the estimating department is, the scale of opportunity available for the future, and other such factors. Quick approximations for *guidance* are all well and good.

It is quite another matter to give such prices when you are running a business, preparing tenders, or seeking accurate financial reporting in any context. In the context of the principles of estimating there is a clear need to avoid the 'guesstimate'.

Let me give you some examples where 'guesstimating' failed.

A major authority used standard rates for outside maintenance work. The work often took place in difficult conditions and over a large geographical area. There was a standard rate to cut down one metre girth trees, leaving the roots intact.

- You may have to work in summer or in winter, in good or in bad ground, and with good or bad access.
- These could be single trees or a group of trees, near to or remote from a public highway.
- The tree could be coniferous or deciduous, you may or may not be able to burn it.

It is extremely unlikely that a single price could sensibly be used for this, or any comparable, operation.

A second example involves a client who intended to buy a local marina and who called into the office one evening seeking help. The attraction for the purchase was the availability of land on which to provide a further mooring basin. The client's problem was that access to the new basin from the adjacent river cut through a public right of way. He would have to provide a footbridge across the access channel to maintain the right of way. He wanted an immediate price for a bridge as the purchase was to be completed the following day. He had no drawings.

I explained that prices could not sensibly be provided without full data and certainly not 'off the cuff'. The client saw the point but nonetheless asked for a price which he would not hold us to and he would also allow us to confirm our price later, and then we could also build the bridge.

Pricing high (or so I thought) I gave a 'present day' price of £32 000.

When we received the drawings, the bridge proved to be constructed of high quality concrete, clad with natural stone and rather intricately shaped. The price was £86 000.

We were embarrassed. The client felt that he had been misled. I had fallen into the trap I recommend others not to fall into.

As a very young Contracts Manager I was asked to price the temporary accesses to the bridges on a bypass. One access alone was 400 metres long and had to accommodate large articulated lorries and 100 tonne mobile cranes, as well as having a Bailey bridge to cross a sizeable river.

My 'present day' price was £1 680 000. The estimator, a person of some considerable experience, felt that £450 000 was more realistic for the work. The lower figure was taken. We won the job and struggled financially as a result.

An excellent construction company, hard working, highly skilled and with a top class reputation, used percentages to cover the site set-up, company overheads and profit. The percentage did vary but a specific figure was used for each different type of work.

Site agents could order whatever they needed and still be charged the same fixed percentage. There was therefore no incentive to save money. The percentage was not worked to and was in reality a 'guesstimate'. The business was struggling.

On one contract, work in a manhole was carried out by:

- isolating the manhole by blocking the pipeflow through the manhole at the adjacent upstream and downstream manholes
- overpumping the existing pipeflow using a 250 mm electric pump
- removing 0.5 m^3 of concrete from the manhole.

The price for breaking out the concrete was £300 per m^3. The company would be paid £150 for the work. There were no other allowances. The estimator explained that £300 was the standard price for breaking out the concrete in manholes. He had 'guesstimated' the figure. A realistic cost for the operation is £2800 based on a minimum two week hire for the pump and hoses. The required rate, based on moving 0.5 m^3, is £5 600 per m^3.

In conclusion I would suggest that a 'guesstimate' can be a helpful indication of price but nothing more. Even an assumed price for ready mixed concrete supply, in error by £1 per m^3, can lead to a substantial error. A pricing error on quite minor items will become considerable on a major job using major quantities.

Do not 'guesstimate' your estimates.

MARKET FORCES

Market forces, while not sources of prices in themselves, can dramatically affect the prices you either receive or provide.

Resource availability is a key factor. Shortages increase prices quickly and surpluses decrease prices quickly. The rates of increase and decrease are greater than those shown in the normal cycle of activity. Estimators need to be aware of this.

The degree of competition affects pricing levels at all stages of the estimating process. Simple jobs using standard materials, concrete reservoirs are an example, tend to be very keenly priced and tender prices vary little. Everyone can do the type of work reasonably well and each tenderer tends to use the services of the same, or very similar, sub-contractors.

Sensible contractors often specialize in a type of activity where they have key skills which the opposition does not possess. River and marine works using floating craft, difficult pipelaying, tunnelling are examples. You are more likely to get sensible prices and sensible margins on this than on routine works. Contractors lacking the expertise to do such work are usually unwise to try to compete with those of proven ability. Select list tendering procedures and the requirements of the *CDM Regulations* for competency should be beneficial to the more specialist contractors.

It can be extremely unwise for the emerging contractor to compete hard for specialist work with a view to beating the specialist. We can all win work, but business is about profit and survival.

Companies tend to work well in areas local to their base. They know the area, the clients, the resource capability and availability. These are the conditions needed to do a successful job. Work in unfamiliar areas can lead to difficulty, dealing with unknown markets or resources.

Estimators need to beware of making assumptions when tendering in unfamiliar areas. I recall a company which excelled in the construction of closed concrete box reservoirs. Invited to tender for two in an unfamiliar rural area, they submitted a high price. Much to their surprise they won the work — and struggled to complete it. They found it difficult to obtain the right resources at the right time — both materials and sub-contractors were a problem. Overall performance was vastly different to work carried out closer to their base.

COMPUTERIZED ESTIMATING

The estimator is responsible for the accurate prediction of construction costs on contracts which have varying degrees of risk. The costs are for work to be carried out at some time in the future, under conditions which the estimator can only imagine, and by people whose capabilities are unknown. There is a large element of the unknown in the process and estimators must assign money on the basis of their own judgement. Whatever methods are used to obtain the price, the estimator's use of personal experience of the work, the art of assessing appropriate resources and the ability to do simple arithmetic, will remain central to the estimating process.

The initial use of slide rules, logarithmic tables and then pocket calculators to help with the arithmetic of estimating has now been overtaken by the use of computers. As most computer systems are approved by people who may know little about them, it is important that the estimator is fully conversant with any proposed developments. In financial terms, £3500 will pay for a competent computer, a printer and the necessary software. A small increase in individual efficiency will soon pay for the provision.

Most current bills are prepared on a computer, turned into paper format and sent to contractors to price. The tenderer prices them manually or enters them into his computer and then prices them. The prices are then put back onto paper and returned to the client who puts them into his computer and checks them. The logic must be that the paper stages are eliminated and the operation remains computerized at all stages.

Many computerized systems are available and all are improving constantly. Intense commercial competition between the various software houses results in constantly improving quality and relatively steady prices. Time saving facilities, e.g. bill scanning and optical character recognition, have improved dramatically.

Some companies have their own in-house system. Such systems can suffer in comparison with modern software if they are not constantly updated. All major construction organizations now use computerized estimating systems for all values and complexities of work. Small bills of quantities, single subject tenders (surfacing and painting are examples), work in small organizations, can often be better priced manually. By and large, however, estimators prefer to stay with a computerized system once they have got used to it.

It is preferable to teach new staff how to estimate manually before moving them onto computer work. However, comparisons show that the growing capabilities of computerized estimating systems have increasing advantages over the standard shortcomings of manual methods.

In a business environment where clients want tenders to be returned more quickly, and where contractors preparing tenders have to provide a bill of quantities, any system which can increase the speed and quality of presentation must be encouraged. Computerized systems do this. Some of the comparative advantages and disadvantages of manual and computerized systems are given below.

Relative merits of manual estimating

Advantages

- The system requires minimal setting up time so the estimator can start pricing quickly.
- An early 'feel' for the tender can be gained by looking at the bill as it is priced.
- There are no computer hardware/software problems.

Disadvantages

- Manual estimating provides limited resource breakdown, usually just totals for labour, plant, materials and sub-contractors. There is no further breakdown (into types of sub-contractor, plant, materials or labour for example).
- Late changes by clients or others are difficult to accommodate.
- The process of producing tenders is usually slower than with a computer system, especially for larger bills of quantities.
- The bill of quantities and preliminaries totals and adjustments are carried out separately. It requires a further stage to bring all the prices together.
- It is difficult to make changes after the tender is finalized. The system for doing this is slow and little time is available.

Relative merits of computerized estimating

Advantages

- A comprehensive breakdown of the labour, plant, materials and sub-contract totals is available.

- The comparison between the various materials and sub-contract quotes can be speedily and simply made.
- Late changes to the bill of quantities can easily be dealt with.
- Changes made at the tender finalization stage can be incorporated into the bill of quantities quickly.
- Systems are generally more speedy and flexible than manual systems.
- The estimator's work can be checked remotely if the relevant computers are linked.

Disadvantages

- Computer breakdown can lead to major difficulties.
- The 'mechanical' system can lead to a lack of 'feel' for the tender.
- Errors can occur—incorrect rates used, items initially unpriced may be omitted for example. Lengthy Quality Assurance systems are needed to avoid this.
- It can be slow and time consuming to set the tender up on the computer before pricing can begin.

Another useful comparison between computerized and manual systems is to see how some of the features of the estimating process are catered for by the two systems. Table 13 gives some guidance on this.

Table 13. Comparison between computerized and manual estimating systems

Feature to be compared	Manual	Computer
Capital cost	None	£3500 per system (approx.)
Ancillary staff required	1 clerical worker for every 3 estimators	None
Staff for enquiries	1 staff member for every 3 estimators	Automated systems form part of some packages which save some time. Allow 1 staff member for every 4 estimators
Entry time onto computer/paper	Copying bill pages onto a pricing sheet or drawing lines onto a copy of the bill—both quick	Typing in quantities is quick but typing in the descriptions is slow. Using a scanner and OCR is very quick although some editing time is required
Pricing time	Intermediate calculations by calculator. Only dealing with money, not resources. Pricing notes required. Resource prices required early in the tender stage. This is a slow process. Write in pencil so that mistakes can be erased	Selection only from a resource library with money not considered at this stage. Opportunity to use previous build-ups of resources and opportunity for automatic pricing, from CESMM codes, for example. Very fast
Pricing resources in rates	No list of resources available. Very time consuming to go back through	Prices for a list of resources can be entered instantly right up to

Table 13.   Comparison between computerized and manual estimating systems (contd)

Feature to be compared	Manual	Computer
	rates changing prices, and global adjustments are often made. Cannot make very late adjustments. Individual rates may therefore not have the money in the right place	the last second. Wastage and conversion rates easily changed. Money ends up in the right place
Tender amendments	Altering of quantities means manual recalculation of whole bill	Quantities altered on screen with automatic recalculation
Summary	Manual calculation by calculator	Touch of a button
Sub-contract and materials comparisons	External manual comparison sheets used. Very slow	Sophisticated comparisons which can update bill and in addition replace L, P, M resources with sub-contract resources. Very quick. Rates can be taken from other sub-contractors in many different ways
Take offs	Done manually on external bill paper	Forms part of some systems. Description libraries for Standard Methods of Measurement enable high quality bills with quantity calculations linked to bill items
Valuations	Manual valuations from bill rates which may not have money in correct place. Slow manual process	Valuation suite forms part of many systems giving an easy means of producing a resourced certificate
Reports	Usually only a basic breakdown of L, P, M, S/C is produced with a total. Not usually presentable to a client without being rewritten	Many reports produced, such as list of priced resources, net bills, selling bills, spreadsheet type report, wastage reports and many others, limited only by imagination. All reports highly presentable and obtainable instantly
Addition of margin	Calculated by hand calculators. Usually only possible to take a broad brush approach as it takes a long time to get a result	Can be added at different rates to different collections of items with result know instantly
Preliminaries book	External hand-calculated sheets with result transferred to a manual summary sheet	Can form part of some systems so that the tender is seamless, transferring result to an automated summary sheet
Rationalization of resources	Not available	Some systems allow quantities to be reconciled, e.g. increasing brick quantities to the next whole thousand or keeping rollers on hire for whole days
Automatic pricing	Not available	Some systems allow the use of CESMM codes to generate automatic build-ups for items which are always priced in a certain way

81

Table 13.   Comparison between computerized and manual estimating systems (contd)

Feature to be compared	Manual	Computer
Previous build-ups	Not available	All systems allow copying of previous build-ups. Some allow items to be grouped together, e.g. all earthworks items can be selected and priced together whenever they occur
Unpriced items	Not available	Unpriced items can be located instantly
Flagged items	Literally 'post-it' notes stuck on pages	Items can be flagged electronically, e.g. to say 'revisit this item!' or 'overmeasured!'
Link to tender programme	Can only be reconciled in an overall way	Individual resources can be totalled for various sections of the works and the programme reconciled in a detailed manner
Storage and transmission	In boxes in archives, usually safe	On hard drives with some risk of loss. Floppy disk 'back-up' can also be provided

6

Getting the final price—the tender and other totals

WORK ITEMS

In Chapter 2 the following elements of price which are used in the estimating process were examined.

labour	shown in estimates as	L
plant		P
materials		M
sub-contractors		S/C

Commencing with the *quoted* cost for each resource we then added various extras (wastage on materials, transport and fuel on plant, travel payment for labour, for example).

Preparing the price to be charged for an item, or the tender sum to be requested, involves the calculation of how much the various resources used actually cost. We need the cost of these items to arrive at the *net* cost of the resources we use. It is net costs which we use in our tender preparation.

The resources, and their net costs, were then used in Chapter 4, where we considered the calculation of rates. Whether we use manual or computerized methods, and computers are an electronic form of the manual system, the effect is the same. Each item of the workscope of the bill of quantities is priced on the basis of the resources which the estimator feels are required to carry out the item.

The sources of the prices we use are varied. The essential requirement is that the estimator uses the ones that are *relevant* to the work being priced and that they are *correct*. Provided that we use *correct* prices the source is immaterial, if we are considering price only.

For future success, someone must agree to do the work, or to provide the materials or sub-contract services, not just at the price but to the specification and the programme, etc. This is why it can be dangerous to use standard price schedules when preparing tenders. Your tender will be an offer to carry out work at prices which suppliers have not yet agreed to do. This clearly exposes your business to risk. You need to be certain that those offering a service will be able to perform if your tender is successful.

Using the *CESMM 3* format for the bill of quantities, the estimator will complete the pricing of Classes B to ZZ inclusive on this basis. In a manual system the calculations will be carried out in a format similar to that used in Chapter 4 (Fig. 4).

In terms of the tender preparation itself we could say that:

> *Stage 1* is to price the work items in the bill of quantities on the basis of the net cost of the resources of L, P, M and S/C used.

Site set-up — preliminaries/site overheads

These were considered in Chapter 3. The estimator uses a standard check-list to decide the resources needed to staff the site, provide accommodation and provide any other items which are necessary to enable the works to be successfully carried out.

The pricing of the site set-up will be in the format of the resources used — labour, plant, materials and sub-contractors. The prices used need to be relevant to the work being priced and to the business (staff salaries, for example). Again, using the *CESMM 3* format for the bill of quantities, the estimator will allocate the sums of money in the various parts of the site set-up price to the relevant items in Class A of the bill.

In terms of the tender preparation itself we could say that:

> *Stage 2* is to price the site set-up items in the bill of quantities on the basis of the net cost of the resources of L, P, M, S/C used.

TRANSFERRING PRICES TO THE BILL OF QUANTITIES

Using a manual system

The estimator has calculated the various rates for use in the bill of quantities, using printed forms as, or similar to, Fig. 4. The bill of quantities is now set out as detailed below.

The bill will usually have the reverse side of each page left blank. When the bill is opened the left-hand page of each double page will be blank and the estimator can enter data as shown below. In the following example, net calculated price is shown as:

Item	Description	L	P	M	S/C
1	Calculated price per unit	6.84	4.56	5.21	16.50

Each left-hand page of the bill is set out as:

Item	L	P	M	S/C	L	P	M	S/C
1	6.84	4.56	5.21	16.50	68.40	45.60	52.10	165.00
	Stage 3 Estimator inserts resource price for a single unit of measure. All bill items are priced and entered into the bill in this manner. The item is that shown in the bill of quantities. The unit cost could be per m³, per m, 1 no., etc.							
					Stage 4 Calculator operator extends the individual rates inserted by the estimator to cover the number of units in the bill. In the case of Item 1, 10 units are assumed to be required by the bill i.e. 10 m³, 10 m etc.			

Stage 5

Once the calculator operator has extended the individual rates put into the bill by the estimator (*Stage 4*), the following steps occur. The calculator operator:

- totals the L, P, M, S/C for each page
- adds page totals to give totals of L, P, M, S/C for each Class or Section of the bill
- adds Class or Section totals to give the Bill total for L, P, M, S/C.

A complete build-up of the resource costs in the tender is now available and will be an excellent tool for contract planning if the tender is successful.

First, however, the estimator must carry out a resource cost check. When pricing the various elements of the work, resources were costed to each element as necessary on the basis of the individual rate requirement. In practice, contracts work on the basis of an overall site provision. In addition, labour and plant is provided on the basis of a daily or a weekly hire. The estimated allowance is likely to be for part of a day(s) or week(s). Sites also tend to operate on steady numbers of operatives with a build-up and run down in numbers at the beginning and end of the job. It is important therefore to check that resources for labour and plant allowed in the tender at this stage conform to what happens in practice. We carry out a Resource Aggregate Tender Check (RATC).

Resource Aggregate Tender Check — Stage 6

This check is carried out by:

- preparing a bar chart programme for the construction work — usually a fairly simple exercise
- resourcing the activities on the programme with the appropriate resources of labour and plant which we envisage will be on site for the various activities
- adding the various resources to obtain programme totals for
  ○ the various types of labour
  ○ the various types of plant
- costing each item of labour and plant and adding them together to obtain the total cost of labour and plant.

  RATC — summing the programmed requirements for labour and plant
  $$\sum L + P$$

Figure 5 gives an example of a costed programme.

- The L and P totals in the bill of quantities (*Stage 5*) are now compared with those from the programme (*Stage 6*). In practice, the totals usually compare quite well. The overall check helps to prove the original pricing approach. The difference between the two totals is generally found to be around 10% in most cases.
- Which figure do we take as correct? The total figure or the programme figure? From my own experience I found that
  ○ On simple contracts (straightforward sewers, small bridges, concrete reservoirs, etc.) the L and P totals in the programme are greater than those in the bill.

Fig. 5. *Wall construction—labour and plant costing*

		MONTH 1				MONTH 2				MONTH 3					MONTH 4			
	WEEK	1	2	3	4	5	6	7	8	9	10	11	12	13	14	15	16	Total
PROGRAMME																		
Excavate																		
Bases																		
Walls																		
Reinstate																		
RESOURCES																		
LABOUR																		
Operatives		4	4	4	4	4	4	4	4	4	4	4	4	4	4	4	4	
Tradesmen			2	2	4	4	4	4	4	4	4	4	4	4	3	2	2	
PLANT																		
Crane		1	1	1	1	1	1	1	1	1	1	1	1	1	1	1	1	
Compressor & Tools		1	1	1	1	1	1	1	1	1	1	1	1	1	1	1	1	
Pump		1	1	1	1	1	1	1	1	1	1	1	1	1	1			
Excavator		1			1	1		1	1				1	1			1	
Dozer					1				1				1			1		
Lorries			2			1				1				1				
Roller					1	1							1			1		
COSTS (£)																		**TOTAL**
LABOUR																		
Labour (£360 per man-week)		1440	1440	1440	1440	1440	1440	1440	1440	1440	1440	1440	1440	1440	1440	1440	1440	23 040
Tradesman (£400)			800	800	1600	1600	1600	1600	1600	1600	1600	1600	1600	1600	1200	800	800	20 400
																		43 800
PLANT																		
Crane (£800)		800	800	800	800	800	800	800	800	800	800	800	800	800	800	800	800	12 800
Compressor & tools (£180)		180	180	180	180	180	180	180	180	180	180	180	180	180	180	180	180	2 880
Pump (£200)		200	200	200	200	200	200	200	200	200	200	200	200	200	200	–	–	2 800
Excavator (£800)		800	–	–	800	800	–	800	800	–	–	–	800	800	–	–	800	6 400
Dozer (£900)					900				900				900			900		3 600
Lorries (£1000)			2000			1000				1000				1000				5 000
Roller (£700)					700	700							700			700		2 800
																		36 280

PLANT TOTAL   £36 280
ADD on/off charges
  Crane   1000
  Compressor & tools   150
  Pump   150
  Excavator   400
  Dozer   600
  Roller   300
TOTAL PLANT COST   £38 880

TOTAL LABOUR COST = £43 440

ADD to the bill L and P totals to bring them to the programme totals.

○ On complicated work (treatment plants, for example) the L and P totals in the bill are greater than those in the programme. LEAVE bill totals as they stand. *Do not* reduce the L and P total in the bill to that of the programme.

In my opinion, the logic for this is that simple jobs tend to run on steady levels of labour and plant. Work gears up and down each day and the simplicity of the work tends to preclude everyone being totally busy all the time. There is an element of cost not picked up by the initial estimate. The overall check corrects this.

However, on complicated jobs there can be a lot of repetitive labours (making holes through walls, for example), which are not picked up by the programme costing system. It is too simplistic. It would therefore be quite wrong to reduce the labour and plant totals.

Using a computerized system

Any price, any tender for work, consists of adding the net costs of all the resources needed to complete that work. This gives us the total net cost and we then include the overhead. To this cost is added an element for profit.

Whatever *system* we employ, the *method* remains the same. The computer provides a system which assists the estimator and fits into the process by:

- supplying historic data on the costs of L, P, M, S/C. This is additional to specific data obtained in reply to quotations sent out for the particular job. It is important that the estimator uses prices which are *relevant* to the particular requirement
- supplying historic data on gang sizes and outputs. This is in support of the practical experience of the estimating team. It is again important that any data used is *relevant* to the particular requirement

It is critically important that any data that you input is *correct*. We are all aware of the phrase 'rubbish in — rubbish out'.

The estimator inputs the bill of quantities data into the computer and adds the resources (and the costs) needed to carry out the work. The computer software processes the data input and outputs in a manner dictated by the software program.

If the software program mirrors the manual system, and ideally it will, then the manual method will be replicated electronically. The printout will be on spreadsheets, of course, and the rates will still need to be input into the bill of quantities. The great benefit now is that the electronic system can perform extremely quickly some of the tasks which were impossibly slow on manual systems.

Arising from the tender vet (this follows in Stage 8), changes will be made to the total price and some of the individual prices. Manual systems took so long to incorporate these changes that the pressure to submit tenders often made accurate changes impossible. Electronic systems make it relatively easy to submit a bill of quantities priced exactly as the contractor wishes.

As with any computer program, you cannot get out more than you put in. To assume that an experienced computer operator with minimal or no practical experience of construction could produce an accurate estimate would be a gross mistake.

The computer is a means of storing information, i.e. standard rates for labour, plant and materials — 'the library' — which may then be grouped into standards, i.e. different gangs for different operations. A problem arises when the estimator has to select, not only the correct gang or build up his own gang, but then has to apply an output which requires a degree of experience. Materials prices can be added at library rates and then adjusted in a particular tender as the actual purchase price is received, but the estimator still has to consider wastage and other factors, which again requires experience.

The estimator has to visualize the construction sequence and methods to appreciate the complexities and elements that need to be addressed. A computer will not do this. Some simple and repetitive operations may be computer estimated by an inexperienced person, once shown which buttons to press, but this does not often occur.

The computer can, and does, save a lot of mechanical calculation work. It is also extremely useful when an operation is run from a number of offices and data can be checked on a remote basis.

Inflation allowance — Stage 7

The estimator has used *current* prices to prepare the tender rates. Completion of the preparation and the extension of the resource elements has given the total resource cost for L, P, M and S/C. The overall check has reviewed and finalized the totals. However, these are still current costs. As work is to be carried out at some future date, an allowance must be made for the effect of inflation.

Evaluation of the cost of inflation was explained in Chapter 3 with reference to the site set-up costs. Assuming that the figures used there still apply, the inflation figure is as follows.

Labour inflation
(% of labour total affected by inflation)

$$\downarrow$$

Inflation = 50% × 5% p.a. ×   £200 000   =   £5000
                    (rate of labour inflation)   (total labour)

Plant
Prices will remain fixed due to nil effect of
inflation   allow   Nil

Materials
£500 000 total
LESS £300 000 fixed price agreements
£200 000 × 5% p.a. × $\frac{1}{2}$ year   £5000
          (inflation rate) (midway through job taken as average)

Sub-contractors
Fixed except for blacktop at £50 000
£50 000 × 4%   £2000
                                                    Inflation total   £12 000

The tender vet — Stage 8

The estimator has prepared and completed the net pricing exercise to this stage working largely alone. Colleagues may be conferred with, the intended construction team will sensibly be in the picture. The site visit will ideally have been a team affair.

It is now essential to have a formal tender vet or check. The number of those attending is dependent on the size and complexity of the job. The basic intention is always, however, to confirm the final price which will be submitted to the client as a tender. The team view will be more valuable than the individual view.

Most companies have a set agenda for tender vets, and this is sensible. The vet does not have to conform to every item on the agenda but all should be aware of the agenda. The intention is to consider *all* relevant points which may affect the price. A detailed agenda will help this to be achieved.

The items likely to be discussed include:

- a full briefing by the estimator describing the job, programme, pricing logic, etc.

- a commercial briefing
- discussion and agreement on
  - labour, allowance and cost
  - plant, adequacy and cost
  - materials, adequate allowance made
  - sub-contractors, adequacy
- programme and resources used
- site set-up
- temporary works issues
- margin (overheads and profit) to be added.

The team vetting the tender ultimately agree the various points. Some agreements have a neutral effect, while some have a positive or negative effect on the net price. The points are totalled financially and the net total of each of the costs for L, P, M and S/C is taken to the net tender total which is then revised accordingly.

PROVISIONAL AND PRIME COST SUMS, DAYWORKS AND CONTINGENCIES

In addition to work which the tenderer prices, the bill of quantities is likely to contain various sums of money inserted on behalf of the client.

- A provisional sum is an allowance for something which may be required—allowing £50 000 for a vehicle turnaround or a building extension would be an example.
- A prime cost sum is to cover a specific requirement, for example allowing £7500 for a Flygt pump of a specified performance and type.
- Dayworks sums are an allowance for incidental items of work.
- Contingency sums are an allowance (10% of the tender total is usual) for just that—contingencies. They are a sign of client prudence and eliminate payment delays when extra costs occur. They also help the client to keep his construction costs within his budget allowance.

OVERHEADS AND PROFIT

Overheads

All companies carry an overhead, a part of their cost which they cannot allocate to contracts on an individual basis. In construction it usually includes head office charges such as:

- directors
- secretariat
- estimating
- purchasing
- accounts
- buildings
- equipment
- auditors
- finance charges, etc.

When the company prepares the annual budget, the overheads are costed as a lump sum, turnover is assessed and the overhead can then be expressed as a percentage of turnover.

$$\% \text{ overhead} = \frac{\text{budgeted overhead cost (£)}}{\text{budgeted turnover (£)}}$$

Clearly, the overhead is a cost which the business has to incur. It must be recovered as a net cost, just as all other costs.

Companies are always reluctant to declare their overhead size for commercial reasons.

Overheads clearly vary from one company to another. In my experience, the overhead varies between 7% and 10%. There has been no apparent difference between large companies and small ones insofar as percentage overhead is concerned. Strong commercial pressure, arising from competition within the industry, forces all organizations to minimize their overhead.

On all tenders we need to add an element to our net price to cover the overhead. For our purposes we will assume that

overhead = 8%.

Profit

Business survives on profit, but competition prevents the addition of a large sum. We will assume that

profit = $2\frac{1}{2}\%$.

TENDER TOTAL FLOW CHART

It is most important that anyone submitting tenders, indeed prices for anything, knows exactly how the price is built-up. While the process has already been explained in some detail, the flow chart (Fig. 6) should help to clarify the process.

Fig. 6.   Tender and price preparation

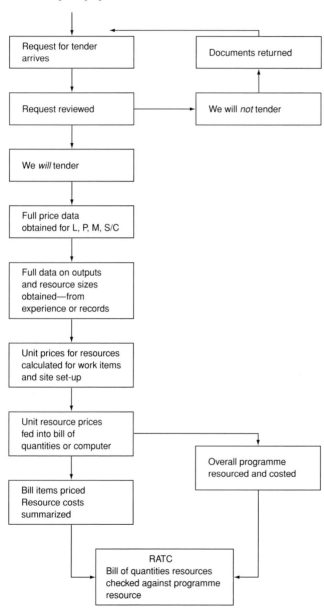

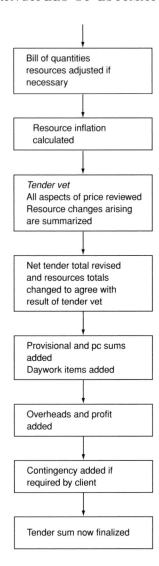

Except for variations of the detail, all tenders or prices are finalized in this way. This is how estimators work. What we now have to do is to ensure that we:

● win the jobs we want to win
● make a profit.

To achieve these objectives we take the *estimate* and use our business logic to finalize our *tender*.

Tender considerations vary from one contract to another, but I believe the following general points apply in most cases.

TENDER CONSIDERATIONS

Competition for work is fierce. We will not often be able to add the 10% to $12^1/_2$% to our net price which is necessary to cover our overheads and provide a profit. A reduced figure will have to be added.

The only way we can then cover our overheads and provide a profit if our tender is accepted is to:

- negotiate all resource prices downwards
- ensure that clients pay *fully* for everything they require.

If our response is inadequate we will ultimately go bankrupt. If our price is too low to cover our resource costs as they occur we need to take action to:

- delay the payments we make. Agreements are set up with materials, plant and sub-contract suppliers which enable us to pay them *after* we are paid by our client
- seek early and enhanced payments from the client. This is achieved by increasing the rates for items of work carried out early in the contract, thus generating a surplus, and reducing the rates for items carried out later by a corresponding amount. The tender total is not changed. This process is intended to give *positive cash flow*. Businesses survive this way. They go bankrupt due to *negative cash flow*.

The bill of quantities is checked for quantities which are likely to increase during the course of the works. Quantities likely to increase will sensibly attract an enhanced rate. By the same token, items likely to decrease in quantity will attract reduced rates.

Work below ground level is likely to change in quantity while items such as concrete, formwork and reinforcement will usually not change. As work below ground will also be carried out early in the contract there is a clear logic in reducing the concrete and similar item rates and enhancing the below ground items.

- Excavation is likely to be sensibly enhanced.
- Small volumes of rock, soft spots etc. are likely to attract extremely high rates.

In Section A of *CESMM 3* provision is made for method-related charges. Tenderers insert charges for such items as they wish. The Additional Description Rules of Section A state that the descriptions for method-related charges shall distinguish between fixed and time-related charges.

An example of this would be for the provision of accommodation.

- Provide accommodation. Fixed charge. Clearly an early item for payment. Enhance it.
- Maintain accommodation. Time-related charge. Claims for extension of time on the contract will benefit from an enhanced sum here.
- Remove accommodation. Fixed charge. The end of the job. Of no benefit to cash flow. Of no benefit in an extension of time claim. Reduce the item.

Clearly, the manipulation of rates will vary from job to job. While professional engineers view the process with a perhaps understandable disdain, the fact remains that if tender prices are too low, all available means must be considered in order to ensure an acceptable contract result. The alternative is for the business to succumb.

When I raise these issues at various seminars, at least one engineer present will tend to make the point that where rates are reasonable and the contract is profitable, then there is no problem with rates for extra work or arguing for money at the expense of the job. I have certainly found this to be the case. The industry would be in a far happier situation if a sensible price were paid for its efforts. It could then concentrate totally on constructing the job. I believe that the *CDM Regulations* and the Latham and Egan reports will greatly assist the problem if they are fully and correctly applied.

VARIATIONS

Any change or variation to the work, is likely to have a financial implication.

If the amount of the work required is reduced then there will probably be a reduction in the value (and the direct cost) of the work. If this occurs the contractor is likely to claim that his plant has become less efficient and that he has recovered less overhead due to the decreased amount of work carried out. This is the *indirect cost* of the variation.

If the amount of work is increased there will probably be an increase in value (and the direct cost) of the work. In addition, the original work may have to be rescheduled and is often disrupted. The cost of any disruption is the *indirect cost* of the variation.

We can say that a Variation has two elements of cost.

- The Direct Cost—this is the cost of carrying out the operation itself. It is usually easy to measure, and easy to present and tends to give rise to few problems.
- The Indirect Cost—the cost of the variation on other activities, usually in the form of disruption and delay to the works themselves. This is difficult to quantify accurately. Cost is difficult to recover as there is nothing to show for it. The indirect cost can be greater than the direct cost of the variation itself.

The timing of the variation will have an effect on cost. A variation of work *before* the work start will lead to some discussion on cost but at least money will not be wasted due to construction operations being interrupted.

A variation *during* the course of the work can have major indirect cost implications. The resources for production will be at their peak so there is a strong likelihood of major indirect cost being incurred.

At the end of a contract, resources on site are at a reduced level. Variations can take longer to carry out as a result. The maintenance of the site set-up for an extended period can have a major indirect cost effect on what is a minor variation with a minor direct cost.

On a properly designed job, variations are rarely a problem. They can be a real problem on 'rush' jobs where the client focus is on other factors (production, retail, finance, etc.). They can also be a real problem when there is a constant flow of variations.

How do we estimate the cost of variations? This will be examined shortly at the end of the chapter.

ADDITIONAL WORKS

Additional works to the *contract* have a similar effect financially to that of a variation. They have a direct cost and an indirect cost. In estimating terms the pricing is exactly the same.

CLAIMS

We will now look at the pricing of variations, additional works and any other events on a contract that are not immediately covered by the accepted bill of quantities—what we call 'claims'.

Many engineers react negatively to the word 'claim'. Some regard claims as attempts to obtain extra money by the contractor and view the claims process negatively. Such reactions not only delay any agreement between the parties, they can also cloud the judgement of those who have to put a value on requests for extra payment. In my experience this often leads to a late and delayed agreement and a level of payment greater than need have been the case.

Most construction contracts have a facility for payment for extra work or unexpected costs and the conditions of the contract should clearly define what is payable. Most engineers would agree to either pay or receive a fair sum.

The problem occurs when the request for extra monies is seen to be too high, or the offer of a sum in settlement is too low. How do we get it right? As we are seeking reimbursement of extra costs the method will be exactly the same as the one we use to prepare an estimate or a tender. We will evaluate the extra resources used and then add overhead and profit.

When we prepare an estimate, however, we are *projecting a future cost* based on the assumed capabilities of the resources we intend to use and their assumed outputs. This is inevitably a theoretical exercise.

When we are receiving, or preparing a request for extra payment, matters are different. The events, outputs, resources, the costs themselves, are not now a matter of theory, they are factual and a matter of record. It should be easy to evaluate a request with the data available and engineers should take a positive approach on this basis.

How should we go about this? Clearly each claim is unique to itself, different from all others. In general terms, however, every contract has similar data which can be used when claims occur.

For example, the construction programme will probably be in bar chart form. It shows what we intended to do. Claims arise when events occur which *prevent* us from doing what we set out to do, leading to programme changes.

We compare the initial programme to the actual 'as constructed' programme. Each activity line affected by the variation will be highlighted and each activity line is a resource activity. From the programme we evaluate delays occurring to each activity line of the bar chart.

Once each activity line and its delay has been considered, the effect on the overall programme is evaluated.

Figure 7 gives an example of a bar chart programme and the activity delays which occurred on each activity line as a result of variations, together with an assessment of the effect on the overall programme.

Fig. 7. Extension of time request

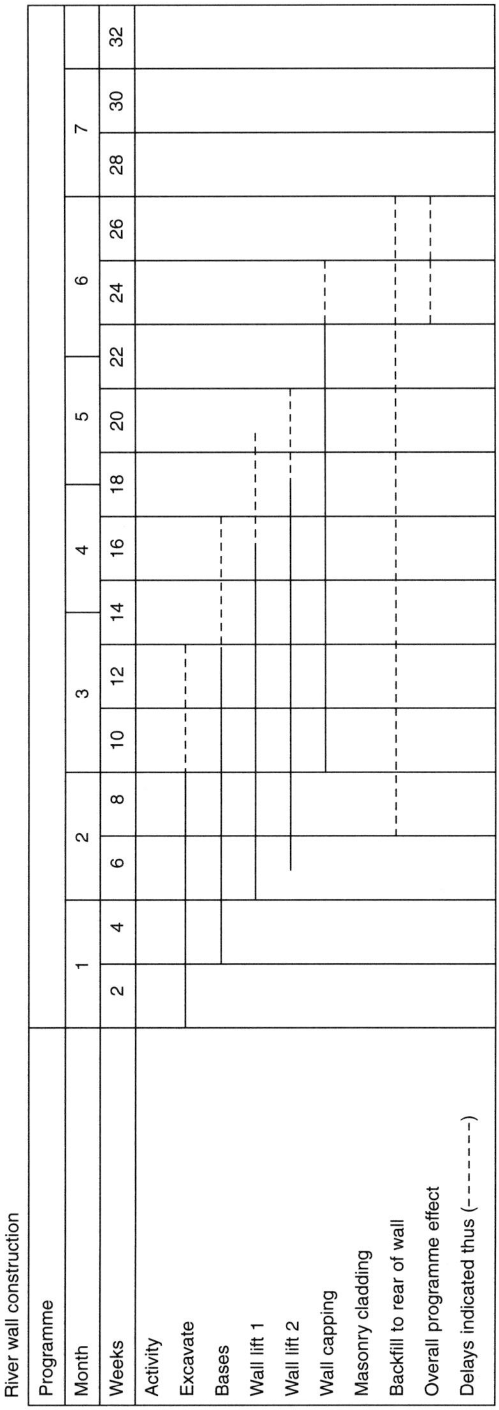

River wall construction

Programme																
Month	1		2		3		4		5		6			7		
Weeks	2	4	6	8	10	12	14	16	18	20	22	24	26	28	30	32
Activity																
Excavate																
Bases																
Wall lift 1																
Wall lift 2																
Wall capping																
Masonry cladding																
Backfill to rear of wall																
Overall programme effect																
Delays indicated thus (– – – – – –)																

99

This gives us the *time effect*.

A good site recording system will tell us not only what went right, but also what went *wrong*. It will record the general effects of the delay, which we have just considered in time terms. It will also indicate the level of resources on site and where they were used.

We should be able to put the resources used on each activity onto the bar chart programme. We can then assess the *additional* resources used on each of the bar chart lines.

The recording system should also note the content of the site set-up. This will be needed to assess the site set-up costs relevant to the variation. We can assess how our resources were affected.

I have indicated that, while the costs of unit resources vary, competition limits the degree of that variation. We should be able to price each resource sensibly from existing data. If you are receiving a claim, use your own cost data initially. This gives us the *cost parameters*.

When we know how resources were affected and can cost those resources, we can evaluate the cost incurred at site level.

We can say

net cost effect $= \sum L + P + M + S/C$ for each programme activity delay at site level

$$+$$

$\sum L + P + M + S/C$ for site set-up overall programme delay
$$= \sum L + P + M + S/C \quad \text{(net cost at site level)}$$

To this we have to add overheads and profit.

Overhead. Every contractor, every business, needs to know the overhead for monitoring and pricing purposes. While good commercial reasons might keep the figure carefully concealed, it is known.

Empirical values quoted tend to give a high percentage overhead. For this reason those submitting claims are likely to quote such figures.

If you are *receiving* a claim, use your own judgement in seeking an agreement.

Profit. Agreement on profit level is needed. It could well be that the percentage profit used is taken from the company financial figures quoted on the Stock Exchange. It seems fair to take an average value.

SUMMARY

In this chapter we have considered the final evaluation of:

- a tender
- a claim, or request for additional payments.

In terms of the principles of estimating, the same method is used for each, we:

- evaluate the required resources of:
  - labour   L
  - plant   P
  - materials   M
  - sub-contractors   S/C
- cost the resources on a net cost basis
- add an overhead and profit.

It is of key importance that anyone pricing work understands this method. Calculation of cost needs to be carried out in a clinical manner and estimators must know exactly what they are doing and how far they have got in the overall process. To ensure this it is necessary to:

- have an accurate bill of quantities (to get accurate prices and to monitor pricing progress)
- develop the price in a series of sections which the estimator can grasp (bite-sized bites).

The process is simple, but it is vital knowledge for anyone aspiring to move to managerial understanding.

Anyone setting up a site needs to be aware of what has been allowed for the various processes.

Anyone on site and in a claim situation needs to understand how to obtain data and, perhaps, prepare the claim itself.

All managers need an understanding of the estimating process in its widest context.

7

Examples of estimates

When I first set out to write about the principles of estimating it was my intention to give examples of a Tender and a Claim. These would be prepared in the manner covered in Chapter 6. It seemed a good idea.

To prepare a Tender, however, would require 'live' documents and a 'live' site. You would have to visit the site before starting to price the work. Such a provision is not realistic. What is *much more important* is that you learn how to price a Tender, or any other work, using First Principles.

The Cost of any Delay which gives rise to a Claim for reimbursement of Extra Cost is calculated by using the same principles which we used when preparing a Tender. However, in this instance we evaluate the Cost (from our records) of the resources *actually* used instead of those we *estimated* we would use when we were pricing future work.

All prices are built up progressively (bite-sized bites) splitting the work into small sections and pricing it section by section.

Before presenting some examples for you to price yourself, I will make some final comments on the costing of resources.

- Labour and Plant are charged to contracts on a daily or weekly basis. They should be priced as such in any estimate
  e.g. for $5^1/_4$ hours — charge 1 day
    33 hours — charge 1 week

- take overall estimates of resource requirements
  e.g. That operation will take one week
      NOT
  80 no $\times$ $^1/_2$ hr = 40 hrs = 1 week

The overall estimate is based on practical experience. The second method is theoretical and can lead to significant errors.

- When working out your costs, work to the nearest pound. Substantial errors can arise when decimal points are introduced early in the process. They also tend to confuse the issue rather than clarifying it.
- Use the approved rate for the classes of labour.
- Charge Plant Hire Rates to the nearest £. This is how they will usually be charged to you. The avoidance of the decimal point will also help to avoid any confusion.

To help provide you with a full understanding of the Estimating process I will treat each of the following examples as an *approximate* Tender exercise and include allowances for the cost of the site set-up, overheads and profit.

Each example given is provided with an answer which is derived manually. You can do your preparation manually or use a computer program if you are practised at it. Answers are given in full to help your understanding.

Remember to work neatly and do the work a step at a time. If you are unsure, take my answer as a guide to get you started. Use a pencil and lined paper. Set sheets out in a similar fashion to those shown in Fig. 4 of Chapter 4.

Each example should be completed in around one hour. This is far quicker than any tender process so it is, of necessity, approximate.

Item	Description	L	P	M	S/C
	Example 1				
	Provide, handle pitch and drive Larssen 20 w piles 8 m long, driven 6 m into clay soil. The length of the pile run is 60 m. Paint piles with 1 coat Lowca black before driving and add 1 further coat to exposed surfaces on completion of driving.				
1	Provide Larssen 20 w piles 60 m × 8 m long = 480 m^2 480 m^2 × 147.2 kg/m^2 = 71 t 71 t @ £550/t for 480 m^2 delivery			39 050.00	(A)
	Therefore 1 m^2	–	–	81.35	–
2	Offload and paint 1 coat Lowca 30 tonne crawler crane on/off site paint both sides — 960 m^2 10 m^2 per hour using 2 men + crane = 100 hours (inc off load) Therefore 2 men + crane × 100 hours = 2 @ £8.00 × 100 (L) = 2 @ £865.00 = (P) Lowca black 100 gallons @ £10.00	1600.00	500.00 1730.00	1000.00	
	Offload and paint 480 m^2 Therefore 1 m^2	1600.00 3.33	2230.00 4.65	1000.00 2.08	(B)
3	Handle, pitch & drive 60 m 600 cfm compressor on/off site gates and timbers for driving on/off piling hammer on/off folding timber wedges etc		300.00 300.00 200.00	100.00	
	Drive 60 m — 6 m/day = 10 days *Labour* 10 days × 9 hours × 3 men @× £8.00 *Plant* 2 weeks × crane @ 865.00 compressor @ 400.00 gates @ 150.00 hammer @ 300.00 1715.00	2160.00	3430.00		
	C/F	2160.00	4230.00	100.00	

Item	Description	L	P	M	S/C
	Handle pitch & drive 480 m² B/F	2160.00	4230.00	100.00	(C)
	Therefore 1m²	4.50	8.81	0.21	
4	Paint exposed piles 60 m × 2 sides × 2 m = 240 m²—1 man 1 week 1 × 50 hrs @ 8.00 Lowca black 25 gallons @ £10.00	400.00		250.00	
	Paint exposed faces per 480 m²	400.00	–	250.00	(D)
	Therefore 1 m²	0.83	–	0.52	
	Labour & Plant Reconciliation Net total prices (A+B+C+D) *Labour* 3 men (incl. working foreman)	4160.00	6460.00	40400.00	–
	× 4 weeks × 50 hours @ £8	4800.00			
	To Labour ADD *Plant* on/off site 30 tonne crane × 4 weeks @ £865 compressors ⎫ hammer ⎬ 2 weeks @ £850 gates ⎭	640.00	1300.00 3460.00 1700.00		
	To Plant ADD Therefore Revised New Work Totals	4800.00	6460 NIL 6460.00	40400.00	
5	*Site Set-up* Working Foreman—extra over labour 4 weeks × 50 hours @ £9 2 no piling men (extra cost) 2 × 4 weeks × 50 hours @ £2 Visiting engineer—1 week Visiting QS—2 days Contracts Manager—1 day	1800.00 800.00 620.00 248.00 220.00			
6	Transit van—4 weeks @ £215 8 no. car days—say 2 weeks		860.00 310.00		
7	Module office/welfare on/off 4 weeks hire @ 100.00 (say)		100.00 400.00		
8	Signs, Safety clothing etc. Telephone—allow			300.00 100.00	
9	Travel time 3 men × 4 weeks × 5 days @ £2.50	150.00			
10	Hardcore stand for crane—50 tonnes JCB Hire		120.00	400.00	
	C/F	3838.00	1790.00	800.00	

Item	Description	L	P	M	S/C
	Net Total Site Set-up B\F	3838.00	1790.00	800.00	
	Final net totals Work	4800.00	6460.00	40,400.00	
	Site Set-up	3838.00	1790.00	800.00	–
	Total net price	8638.00	8250.00	41,200.00	
	ADD Insurance (allow for excesses) — 2%	1162			
	ADD Inflation	NIL			
	ADD Contingency	NIL			
	ADD Overheads, Profit 20% of £59,250	11,850			
	Tender Total =	71,100			

Example 2
20 m of a river-bank has slipped and is to
be reinstated using stone-filled gabions.
Material excavated for the gabions is to be
put into the bank behind the gabions.
Section through river-bank

SLUMP   WATER 1 m deep

Gabion Requirement

STONE BED

2 m × 2 m × 1 m Gabions

Item	Description	L	P	M	S/C
	Quantities				
	Excavation 60 m^3				
	Gabions 50 no. × £60.00				
	Stone Fill £10 per tonne				
	Method				
	Excavate slipped material and place				
	gabions immediately		200.00		
1	Long arm excavator on/off site				
				3000.00	
2	Provide materials				
	50 no. gabions × £60.00				
	stone fill 50 no. × cm @ 2 t/cm				
	= 200 tonnes + fill to bed of river (say				
	50 tonnes)—250 tonnes @ £10			2500.00	
3	Install gabions 6 no. per day including				
	excavation—8$^1/_4$ days total add back				
	fill and tidy site therefore 10 days total				
	10 days × 9 hour—2 weeks		1880.00		
	2 weeks Komatsu excavator at £940	2160.00			
	10 days × 9 hours × 3 men × £8.00				
	Net work total	2160.00	2080.00	5500.00	
	Labour and Plant Reconciliation	2160.00			
	Labour total 3 × 9 hours × 10 days × £8		1880.00		
	Plant Komatsu × 2 weeks × £960		200.00		
	on/off site	2160.00	2080.00		–
	To labour and plant				
	ADD—NIL	2160.00	2080.00	5500.00	–
	Therefore revised net work totals				
	Site Set-up	248.00			
4	Visiting staff 2 days @ £124		430.00		
5	Site van—2 weeks × £215		40.00		
	2 no. car days @ say £20		130.00		
6	Portable toilet on/off + 2 weeks hire		100.00	150.00	
7	Signs, safety, allow			50.00	
8	Expenses (phone calls, etc.)	60.00			
9	Travel time 3 men × 10 days × £2.00	308.00	700.00	200.00	–
	Net Total Site-set-up	2160.00	2080.00	5500.00	
	Final Net totals Work Wash	308.00	700.00	200.00	
	Site Set-up				
	Total net price	2468.00	2780.00	5700.00	

Item	Description	L	P	M	S/C
	ADD Insurances (allow for excesses) — 2%	220.00			
	ADD Inflation	NIL			
	ADD Contingency — 5%	550.00			
	ADD Overhead and Profit — 20%	2340.00			
	Tender Total	14058.00			
	Example 3				
	A circus visit to the local park has just ended and we have to reinstate the area.				
	Over an area of 3 hectares (30 000 m^2), 50% of the 150 mm thick topsoil has been lost, the remainder is rather a mess. All grass needs replacing and 50 shrubs have been lost.				
	Requirement				
1	Strip topsoil remaining — 30 000 m^2				
2	Harrow subsoil and bring to a fine tilth				
3	Replace existing topsoil and provide 50% new				
4	Harrow and seed				
5	Replace shrubs				
	Quantities				
	topsoil provision				
	30 000 m^2 × .075				
	= 2250 m^3 @ £10 per m^3				
	shrubs — £5 each				
	harrow and seed — 20 p per m^2 S/C				
	Method				
1	Strip remaining topsoil and store at edge of site				
	30 000 m^2 × .075 = 2250 m^3				
	955 Cat with 3 in 1 bucket can strip at rate of 20 m^3 per hour =				
	112.5 hours — SAY 15 days				
	on/off site		250.00		
	15 days =3 weeks @ £770.00		2310.00		
	Banksmen — 15 days × 8 hours @ £8.00	960.00			
	Strip topsoil 30 000 m^2	960.00	2560.00	–	(A)
	Therefore 1 m^2	.03	.09	–	–
2	Harrow subsoil and bring to tilth — S/C farmer + tractor — £350/day allow 2 days				700.00
	per 30 000 m^2	–	–	–	700.00(B)
	Therefore 1 m^2	–	–	–	.02

Item	Description	L	P	M	S/C
3	Provide replacement topsoil 30 000 m² × .075 = 2250 m³ @ £10			22,500.00	
	Spread existing and replacement topsoil Volume = 2250 + 2250 — 4500 m³ 955 Traxcavator spread and backblade at 15 m³ per hour				
	– 300 hours				
	– 2 machines — 100 hours per week				
	– takes 3 weeks				
	– 2 no. machines on/off		250.00		
	– 2 no. × 955 Cats × 3 weeks @ £770.00		4620.00		
	– 2 no. banksmen × 3 weeks @ 50 hours × £8	2400.00			
	per 30 000 m²	2400.00	4870.00	22 500.00	(C)
	per m²	.08	.16	.75	
4	Harrow and seed				(D)
	S/C 30 000 m² @ 20 p				6,000.00
	per m²				.20
5	Replace 50 no. shrubs — supply 2 no. men × 10 hours @ £8.00	160.00			
	50 no. shrubs			250.00	
	per 50 no. shrubs	160.00		250.00	(E)
	per 1 no.	3.20		5.00	
	Final net Work total (A+B+C+D+E)	3520.00	7430.00	22 750.00	6700.00
	Labour and Plant Reconciliation Labour — parks department staff who will allocate to other duties. Allow	NIL			
	Plant — charged for full weeks at hire rate. Allow		NIL		
	Revised net Work total	3520.00	7430.00	22 750.00	6700.00
	Site Set-up We will use 2 × 955 Traxcavators to do initial soil strip. This will take 1½ weeks, 4½ weeks overall programme				
6	Working foreman on site 5 weeks × 50 hours @ £9	2250.00			
7	Visiting staff 5 days @ £124	620.00			
8	5 no car days — 1 week		125.00		
9	Mobile office/welfare on/off		100.00		
	5 weeks hire @ £100		500.00		
10	Signs, clothing, safety — run			70.00	
11	Travel time 9 man weeks × 5 days @ £1	45.00			

Item	Description	L	P	M	S/C
	Net Site Set-up Cost B/F	2915.00	725.00	70.00	–
	Final net totals Work	3520.00	7430.00	22750.00	6700.00
	Site Set-up	2915.00	725.00	70.00	–
	Total net price	6435	8155	22820	6700
	ADD Insurance (incl excess) 2%	900			
	ADD Inflation	NIL			
	ADD Contingency 2 days L&P	145	600		
	ADD overheads and profit—10%	4575			
	Tender Total	50330			
	Example 4				
	Figure 8 shows the plan view of a sewer in roads in a built-up area. Our job is to lay the sewer. The roads are all 12 m wide with an additional 2 m wide footpath at each side of each road. The work is to be carried out under a road closure order so we have no traffic to worry about				
	Requirement				
1	Lay new 600 mm dia sewer between manholes 1 and 6				
2	Construct 6 no manholes				
	The sewer is 3 m deep to invert. All excavated material is to go to tip. The nearest tip is 4 miles away. All trenches and manholes are to be backfilled with type 2 stone.				

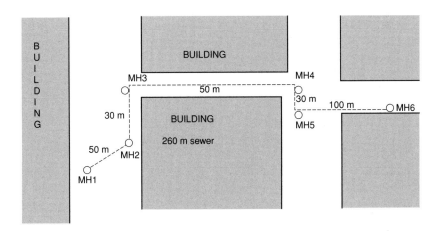

Fig. 8.   Sewer plan

Item	Description	L	P	M	S/C
	Quantities 600 dia. concrete pipe—260 m @ £27.10 – type 2 stone—£6.89 per tonne – precast concrete manholes—£450 complete – HD manhole covers £130 each – tipping fees—£9 per m³ – pipe bedding—£8.00 per tonne				
1	*Programme* Set up site, divert traffic, close roads, all plant to site	Sewer 3 days	Reinstate services etc.		Manholes 3 days
2	Excavate manholes 1–6 incl. (by pipelay gang)				6 days
3	Construct manholes 1–6 incl. × 3 days each				18 days
4	Lay 260 m pipes @ 10 m/day	26 days			
5	Reinstate trenches & manholes—25 m per day		12 days		
6	Clear site		3 days		
7	Expose services ahead of trench – 26 domestic properties @ 2 hours each = 52 hours – mains services 16 no. × 3 hours = 48 hours				
	Total Hours = 100		10 days		
	Gang sizes—Pipelay L P 2 no. on trench supports 1 no. on pipelay/bedding 1 no. on banksman 4 no. × 9 hours × 5 days @ £8.00 Weekly cost	<u>1440.00</u>			

	Plant
	On/off site 250 Komatsu weekly hire
	          100 compressor
	              tools
	       50 compactor
	      300 trench supports
	       50 2" pump
	       50 4" pump
	       80 10 KVA generator
	          lorry (24 t)
	
	        <u>Weekly hire</u>
	<u>£880</u> On/off site
			940
			130
			30
			100
			300
			60
			100
			110
			1115
			
			<u>2885</u>		

Item	Description	L	P	M	S/C
	Manholes				
	L　　　　　　　　P				
	2 men × 50 hours @ £8				
	Weekly cost =	800			
	On/off site — JCB 3 C		660		
	50　　2" pump		60		
	100　trench supports		100		
	50　　compactor		100		
	20　　concrete vibrator		55		
	On/off　　220　Weekly hire		975		
	Expose services ahead of pipelaying				
	260 m of pipework — 10 m frontage to properties — 26 properties each side of road.				
	Domestic, allow 26 no. × 2 hours				
	= 52 hours				
	Main supplies 4 crossings × 4 no. ×				
	3 hours　　　= 48 hours				
	Total　　　　= 100 hours				
	L　　　　　　　P				
	2 men × 50 hours @ £8				
	Weekly Cost	800.00			
	On/off site　50　　compressor		130		
	tools		30		
	10　Tarmac		10		
	cutters				
	40　barriers		50		
	On/off site　100 Weekly hire		220		
	Reinstatement gang				
	L　　　　　　　P				
	2 men × 50 hours @ £8　Weekly cost	800			
	compressor		130		
	tools		30		
	Tarmac cutters		10		
	barriers		50		
	roller		100		
	weekly cost		320.00		
	On/off		130.00		

Item	Description	L	P	M	S/C
	Method				
1	Set up site bring all plant to site and set up barriers etc.				
	Plant on/off site—pipelay gang		880		
	manhole gang		220		
	services		100		
	reinstatement		130		
	pipelay & manhole gang				
	6 men × 10 hours × 3 days @ £8/offload and string out pipes	1440			
	allow mobile crane 3 days @ £240		720		
		1440.00	2050		(A)
2	*Lay pipes*				
	• 21 + 6 = 32 days—$6^2/_5$ weeks				
	$6^2/_5$ @ 1440.00(L) 2885.00 (P)	9216	18464		
	• pipes—260 m + $2^1/_2$% waste @ £27.10			7222	
	• pipe bedding (1 metre wide trench)				
	260 m × 1 × .35 × 2 t/m³ @ £8.00			1456	
	• trench dig to tip 780 m³ @ £9			7020	
	• type 2 fill				
	260 × 1 × 2.5 × 2 t/m³ @ £6.89			8957	
	Lay pipes	9216	18464	24,655	(B)
	per m	35.45	71.01	94.83	
3	*Construct manholes*				
	• 6 no. manholes and covers @ £580			3480	
	• concrete surround to manholes				
	6 no. × 1.5 m³ × £60			540	
	• construct 6 no. × 3 days = 3.6 weeks				
	3.6 weeks @ 800 (L) 975 (P)	2880	3510		
	Allow				
	• type 2 backfill to manholes			165	
	6 no. × 2 m³ × 2 t/cm @ £6.89				
	6 no. manholes	2880	3510	4185	(C)
	per manhole	480	585	697.50	
	Expose services				
	2 weeks @ 800 (L) pw 220 (P) pw	1600	440		(D)
	Reinstatement				
4	2 weeks @ 800.00 (L) pw 320.00 (P) pw	1600	640		
	Bitmac				
	260 m × 1.5 m av. × .25 thick				
5	= 97.5 m³ = 195 t @ £30.00			5850	
		1600	640	5850	(E)

Item	Description	L	P	M	S/C
	Total Net Price (A+B+C+D+E)	16 736	25 104	34 690	
	Reconciliation of Labour & Plant Labour net allowed £16736 = 41.84 man weeks – increase to 42 man—weeks ADD	66			
	Plant • pipelay gang 6.4 weeks allowed ADD 0.6 weeks @ £2885 • manhole gang 3.6 weeks allowed ADD 0.4 weeks all plant except JCB		1731 126		
	Revised net L & P < M	16 802	26 961	34 690	
	Site Set-up Costs (Preliminaries)				
6	Foreman 8 weeks @ £870 Engineer 5 weeks @ £620 Quantity surveyor 2 weeks @ £620 Contracts manager 1 week @ £1120	6960 3100 1240 1120			
7	Small cars—15 weeks @ £125 medium car—1 week @ £170 transit van—8 weeks @ £215		1875 170 1720		
8	*2 × 5 m offices* on/off site @ £80 2 × 8 weeks hire @ £45 1 no small toilet on/off site 1 × 8 weeks hire @ £65		160 720 60 520		
9	*Health and safety* 12 sets clothing @ £50.00 signs allow training—24 hours @ £8 first aid allow barrier —100 m × 10 weeks @ £0.25	 192		600 200 100 250	
10	*Site compound* 100 m × 2.4 m hoarding @ £22/m 1 pair gates transport on/off site			2200 250 160	
11	*Services* mobile phones 15 man weeks @ £10 calor gas 8 weeks @ £15			150 120	
12	*Labour charges* 38 man – weeks × 5 days @ £1.50 (Travelling) 38 man weeks × 5 days × 10 hours @ £1.30 (pipe gang premium rate)	285 2470			
	C/F	15 367	5 225	4 030	

Item	Description	L	P	M	S/C
	B/F	15,367	5,225	4,030	
13	*Small tools* shovels, picks etc.				
	5% of labour = £750 ALLOW			750	
14	*Setting out equipment*				
	level, staff 8 weeks × £35		280		
	timber, paint & pins			100	
	pipe laser		500		
	tapes etc.			100	
15	*Signs*				
	small company signs only			300	
16	*Clear site*				
	2 men + lorry × 2 days	320	450		
	tip fees			100	
	Net Site Set-up Cost	15687	6455	5380	
	Final net totals Work price	16802	26961	34690	–
	Site Set-up	15687	6455	5380	–
	Total net price	32489	33416	40070	
	ADD Insurance 0.8% + £500 SAY	1300			
	ADD Bond £1 per cent of 10% p.a.	100			
	ADD Inflation	NIL			
	ADD Contingency	NIL			
	ADD overheads and profit 10% SAY	10000.00			
	Tender Total	117,375			

You should by now be developing an understanding of how Estimates of Cost are prepared. Whatever the purpose of the estimate, the procedure for preparation is the same. Practice the procedure on the types of work with which you are familiar. An ability to Estimate Costs of all types will stand you in good stead.

The layout used for the examples is adapted to fit the page size of the book. You will be using your standard company format which will be different to the one I have used. The principle will be the same though.

You will probably have had some difficulty in adding the numbers. Do not worry; this is quite usual. It does, however, reinforce the need to use pencil and an orderly layout as I have stressed several times previously. There will be no chance of finding errors if your work is untidy, nor will others be able to check your work.

8

Cost, value and budgets

Rightly or wrongly, business is about profit. If we make a profit, the business will survive. If we make a loss, it will ultimately perish. While this may seem obvious, the fact remains that many engineers are uncertain as to how to cost things and to use costing systems in a business-like manner which they fully understand.

My purpose in writing '*Principles of estimating*' has been to provide a simple method of evaluating different types of cost in a way which all engineers should understand. If work is priced on a schedule of rates basis, a simple process of multiplication (quantity x rate) will give us a sum of money which may, or may not, win the job. By pricing the same work on a resource cost basis you know *exactly* what resources have been allowed in the price. Practical experience will help to ensure that you allow the right amount of resource for the appropriate time. When you complete a tender on this basis you will feel confident in your price and your colleagues and senior managers will be able to check your price with confidence.

When you price and win a tender, or gain *any* kind of work on the basis of resource costing, further benefits can be obtained. These benefits are of enormous value to the business, and will be considered below.

Those carrying out the work can plan on the basis of the resources allowed in the price. They do not have to reinvent the wheel and start again. Any tender errors (or benefits) can be picked up before they affect the site and work can be preplanned to mitigate any problems. Work programmes can be resourced and costed, then compared to the tender evaluation. You are aware of the expectation from the outset and know how to meet that expectation. This is far better than to know the expectation but be unsure of whether you will meet it. Diligent application to the work ethic is fine. Blind diligence is foolish.

The process we use to *estimate* costs in our bid to win work can be used to provide *actual costs* which will help us to monitor the business.

We are now going to look at cost, value and budgets to see how a business can be run using quite simple concepts. The process is standard practice in all businesses.

COST

Cost is an accountancy function. It is a statement of what we have spent. The monitoring of cost is a vital part of monitoring the health of the business. While the results are provided by the accounts department, the process is exactly the same as running your own finances. You record what you spend. This is not rocket science. It is a simple process.

Good businesses, with strong survival instincts, monitor costs regularly. Construction businesses traditionally monitor on a monthly basis in phase with the valuation process.

You can monitor whenever it suits you. Good businesses report at monthly intervals or more frequently. When reporting times become more extended, it is often a sign that there is a problem in the business.

A further important point on the frequency of cost reporting is that the costs reported are historic ones. As reporting periods increase there is progressively less opportunity for management to react to any adverse trend. It will be too late!

The best businesses report costs *promptly*, on a *monthly* basis. The costs should ideally be available for any month by around the end of the first week of the following month.

THE CONTRACTOR'S COSTS

Each business allocates its costs in its own way. It allocates to what we call 'cost centres'. However, cost centres are purely a matter of detail — the costs remain the costs.

We have site costs and office costs.

Site costs

Each site has its own monthly costs, split down into the cost centres which best suit the company. An example would be:

- labour   staff salaries
  self-employed labour
  directly-employed labour
  overtime
  travel time
  holidays with pay, etc.

- plant   cars — hire/maintenance
             fuel
             repairs
             insurance

      vans — hire/maintenance
             fuel
             repairs
             insurance

      huts

      own plant   (non-operated)   compressors
                                                 dumpers
                                                 pumps, etc.

      own plant   (operated)   lorries
                                             excavators
                                             cranes, etc.

      hired plant   (non-operated)

      hired plant   (operated)

- materials   concrete   C15
                          C30
      reinforcement — by diameter
      bricks
      blocks, etc.

- sub-contractors   formwork
                        drainage
                        reinforcement, etc.

Your business may also elect to have a further cost type. For example:

- sundries   petty cash
                  stationery
                  telephones
                  electricity.

The following points should be borne in mind.

- No matter how you split them down, the costs remain the costs. The same money is simply being put into different boxes.
- Splitting down the costs into the sundry cost centres makes monitoring easier.
- Do not make your system of cost centres too complicated. This can be demotivating. The system becomes the objective.

Office costs

Each department has its own costs and these will be presented in the same format as the site costs. For example:

departments   accounts
                estimating
                plant
                purchasing
                secretarial
                directors
                buildings
                sundries, etc.

Department costs will consist largely of staff and their associated costs, including cars. Building costs would include depreciation, repairs, rentals, etc. Sundries could be cleaning, welfare, furnishings and similar items.

When we use a site and department costing system such as this, various benefits are obtained.

- It is easier to isolate problems and then to deal with them than when all costs are put in one overall statement. Monitoring is easier.
- Site and departmental managers can take responsibility for their own costs. Management, teamwork, 'ownership' of the business, are all strengthened.
- The amount of delegation in the business is increased and this usually increases the positiveness of the team response. The elements of the business become their own.

THE CONSULTANT'S COSTS

Consultants have sites and head offices just like contractors. However, whereas the contractor has plant, material and sub-contractor costs on site, the consultant's costs are largely for the provision of staff and the buildings which house them.

The costs for a typical small/medium consultancy are shown in Table 14. The costs are split into cost centres and cover a full financial year of twelve months. They are annual costs.

While the costs cover the entire business unit, it is sensible to split them down into sub-units—each site, design team, etc. could have its own costs. This would be logical and make sound business sense.

Table 14. Typical consultancy costs

White Rose Design Ltd	
Annual income and costs	£
Fees received	3 000 000
Staff salaries (technical and management)	1 440 000
Staff salaries (clerical)	170 000
Employers NI	160 000
Health insurance (senior staff)	30 000
Training	18 000
Subsistence and site allowances	42 000
Legal and professional	12 000
Accountancy	30 000
Rent, rates and services	260 000
Telephone and fax	23 000
Light and heat	18 000
Repairs	18 000
Cleaning	12 500
Depreciation, fixtures and fittings	24 000
Equipment lease charges	34 000
Printing and office supplies	55 000
Advertising	10 000
Motor and travel	230 000
Depreciation motor vehicles	30 000
Bad debts	7500
Insurances	132 000
Entertainment	10 000
General expenses	33 000
Staff welfare	12 000
Bank charges and interest	39 000
Total cost	2 850 000
Profit and tax	150 000

From Table 14 you will see that:

- staff and associated costs total £2 114 000—this is 74% of the total cost incurred
- buildings and maintenance costs total £416 000, some 15% of total cost

the above costs relate directly to the number of staff employed

- general business costs total £320 000 (insurances, bank charges, accountancy, general expenses, etc.). This is 11% of total cost.

VALUE

Value, in business terms, is what things are worth, what people are prepared to pay for what we are selling. It is our income. For a *contracting organization*, value is essentially derived from the following areas.

Contract valuations

Admeasurement contracts (those which are remeasured as work progresses) are usually valued on a monthly basis which is in line with the costing system. In general terms, this allows a direct comparison of monthly cost against monthly value. Other types of contract also tend to be geared to producing a similar monthly value, albeit possibly by different methods.

On a contract which is running normally, that is without disruption, this is usually sufficient to cover the site costs and assist towards covering the organization's overhead. Most contracts do run normally.

Full realization of value is the key to achieving a positive cash flow.

Contract settlements (payments of claims)

These are payments to a contractor for 'loss and expense' in carrying out the work. The loss and expense has already occurred and is reflected in the current contract costs.

Settlements, when made, are therefore largely positive. They are *value* with little *current* cost to offset them. This is clearly beneficial to the contractor. In a highly commercial climate they are vital.

Internal profits

It is possible to reduce overheads by making internal profits or surpluses. While all overhead departments need to work in line with the budget (i.e. not to incur an internal loss), it is possible to improve matters in a variety of ways.

- The plant department can be set an increased profitability target. Using less plant more often, perhaps hiring unused plant externally, generally increasing utilizations, all help. Good maintenance can reduce downtime and extend plant life.

- The purchasing department has to match the prices for materials which were used in the tender. A target can be set to reduce the *order* prices for materials to, say $2^{1}/_{2}$ or 5% less than the prices quoted in the *tender*. On a contract with a material content of 30%, such a saving can reduce company costs by 1% if it is applied across all purchases.
- Set a target for the estimating department to produce more *profitable* turnover than has been budgeted for. Unprofitable work, turnover for the sake of turnover, produces no sensible benefit unless you are desperate for cash. You can pay dearly for this when the jobs you win in this way start to lose money.

In a *consultancy organization* the income, the value in the business, is derived from *fees*. Fees are paid by clients for consultancy work carried out and are payable as a lump sum, a percentage of the job value, or on a time basis.

While figures vary from one job to another, an average fee of 5% of the contract value is a reasonable average to take.

The consultancy considered in this example has

- income of £3 000 000
- costs of £2 850 000
- this gives a surplus of £150 000 per annum
- with 5% fees, the value of work carried out to generate £3 000 000 fee income is £60 000 000.

If we consider the costs of the consultancy

- 75% of costs are staff related, these will be paid monthly on a regular basis
- 15% of the costs relate to buildings, these are directly related to the number of people employed.

Monthly costs are high and regular and ideally need similarly regular income. To maintain a financially stable position we need to consider value carefully. What can we do?

- An extra 1% payable on fees is an increase of £600 000 on the current fees payable. It is sufficient to increase the surplus by a factor of 5 (£150 000 increases to £750 000).
  Strong *marketing* action to achieve a better level of fee seems a reasonable action to take.
- The cost of training in a people business is often very low. It is good

practice to train people to achieve their full capability. This will also tend to help teamworking and enhance the business as a whole.

CASH FLOW

Cash flow in a business is what it says it is — the money flowing through the business. Money flows in — and money flows out.

- *Positive cash flow* occurs when money flows in faster than it flows out. There is an increasing surplus at the bank. It is good practice to have a positive cash flow.
- *Negative cash flow* is the opposite. Here money flows out faster than it flows in. The bank position steadily worsens. It is poor practice to have negative cash flows.

The simplistic answer to cash flow problems is to ensure that you get paid yourself *before* you pay others.

All businesses strive to achieve positive cash flow.

How do we achieve this in the construction industry?
The *consultancy* may consider the following options.

- Decrease direct employment and introduce an element of sub-contracting. Payment will be on a result and not a time basis. Payments should occur after receipt of fees and arrangements should be such as to ensure this.
- Consider an element of homeworking. This is a growing concept. It will ultimately lead to smaller office costs.
- Adopt a strong marketing policy to realize higher margin work.
- Diversify into other areas of work.
- Ensure that you employ the best, most able, people. The increased output they provide will enable a greater work throughput to be achieved.
- Arrange payments from clients pre-contract start (ensure that money flows in *before* it flows out).
- Arrange for some work to be carried out on a cost plus basis (usually paid as time spent on the work).

The *contractor* may consider the options given below.

- Increase the amount of sub-contract work carried out and do such work on a basis which enables payment to be received *before* payments are made.

- Ensure that materials are paid for *after* the client has paid the relevant account.
- Ensure early payments by adjusting the bill of quantities rates (enhancing rates for early payment items).
- Decline to chase turnover at low margins.
- Pursue claims for reimbursement of loss and expense.
- Diversify into better margin work. Negotiated contracts (cost plus work, specialist maintenance, are examples) rather than competitive tendering.

PROFIT

For any business

Profit = income − expenditure
OR
value − cost

It is achieved by working hard to:

- minimize cost
- maximize value.

This is how you will make a profit. You will *not* make it by considering it in isolation. Profit is the result of good business actions elsewhere.

BUDGETS

In this chapter we are considering the financial results which arise from our business activity. While we are particularly interested in construction, all industries and businesses are interested in the results which are the various proportions of

- cost
- value
- profit
- cash flow.

At the reporting periods we work to in our organization we can see how we are faring in our endeavours.

However, by using this system and looking no further we will actually only be considering past performance. Even the most recent costs and values are now historic. To manage a business properly we need to be able to look forward. We do this by producing a *budget*.

A budget could cover a household, a company, a country, indeed any economy. The time-span which the budget covers is variable. A major unit will have a five or ten-year outline plan within which detailed annual budgets will be provided. Companies use annual budgets which match their financial year—January to December and April to March are the usual ones.

We will assume we have an annual budget to cover a twelve-month period.

How do we set the budget? What does a budget consist of?

In this case we will take a twelve-month period—say January to December. We will then predict the cost and value for our business for each month of that period. It is very similar to estimating the price of the construction work itself, but with one major difference. When we were estimating the cost of the resources needed for construction work we relied on our practical experience of similar activities. In the budget exercise we have the past financial results of each part of the business, the past costs and values. We have hard facts to work on.

Setting the budget

While every organization will be slightly different in detail, the principles remain the same. We want to know how the business will perform

- during the next financial year
- on a monthly basis
- in terms of total value − total cost = profit.

While the *overall* result is the final requirement, it is clearly beneficial to build the figures up using a series of cost centres. We can also look at each cost centre to locate particular problems if we do get an excess cost.

For the contractor the budget will cover each site and each department.

The consultant may have an annual budget similar to that shown in Table 14 or may also sensibly split the total into sites and departments.

Using the concept of cost centres increases accountability (each centre has its own budget to work to) and makes the business much easier to manage (the alternative is to manage one large mass). Teams are encouraged to develop and leadership opportunities increase.

The actions taken in our business can be summarized as those shown in Table 15.

Table 15.   Preparing the budget—contractors and consultants

	Contractor	Consultant
Assess value for year	• Take each contract and assess its turnover (in line with tender, usually, and based on the programme). • Consider possible claim settlements. • Assess any internal surpluses based on previous years' experience for purchasing, plant depts, etc. • Decide the likely level of new turnover.	• Take existing work and assess fee income. • Decide the likely level of new turnover and fee income.
Total value =	The total is likely to be of the order of—previous value (turnover last year) + 10%. The 10% addition covers inflation and allows a reasonable growth in the business.	Fee income as last year +10%.
Assess cost for year	• Take each current contract and assess costs. Use earlier contract costs as a guide. Allow for any increased costs (staff salaries for example). • Take each internal department and assess cost (use previous year as a guide). • Add all costs to give total current costs. ADD ○ any cost change (+ or −) ○ inflation allowance ○ increases in cost incurred in new business. (Take new business and costs as allowed in respective tenders.)	• Take the overall business cost or each section cost. Use current costs for the unit(s). ADD ○ any cost change (+ or −) ○ inflation ○ increases in cost to cover new business.
Total cost =	Likely to be previous cost + 10% (say)	Previous cost + 10% (say)
• Assess Total Surplus • Is it adequate? (Will we make a profit? Are there any problem areas?) • Directors agree the final budget.	Total Value—Total Cost • If it *is* fine, we can adopt the budget as it stands. • If it *is not*, where is the problem? What can we do to eliminate it? Agree corrective actions. Adjust budget to conform to revised agreement.	Total Value—Total Cost Act in a similar manner to the contractor

127

Working to the budget

Whatever business you are in, you will now have to work to the agreed budget. The budget could be a single one for an entire business, or it could be one of a series of departmental or site budgets.

Initially, the budget will be a single statement of figures giving an acceptable annual result. We now have to work to achieve that result and to ensure success we need to monitor ourselves regularly against the requirement.

As stated above, the logical monitoring period is one month and most businesses work to this. The annual budget is split down into twelve increments of one month. As we have 52 weeks in one year, the reporting periods are usually of 4, 4 and 5 week periods to give 13 weeks per quarter year. These are repeated to give the 52 week overall total.

The budget could be the same for each month but in construction this is unlikely.

A *contractor* will tend to increase turnover and profitability in summer. The year may start with a low level of work and this may increase later. Or the reverse could apply. It depends on the business cycle of your organization. Contract settlements will have intermittent effects. It is best to avoid these in your budget preparation as they cannot always be relied upon. Equally, if your business is in difficulty, their effect could be crucial.

A *consultant* will tend to have less monthly variation as the turnover is largely based on the number of people employed. Certainly the costs are likely to be steady on a month by month basis.

Income from fees will vary from month to month. In a tight market the results, in cash terms, could easily vary from plus to minus as a result. Internal adjustments can correct this but the cash situation clearly needs careful attention.

In general terms, it is sensible for any business to delay capital spend until later in the year when you will know how close you are to achieving your budget. Early capital spend can lead to deep regrets later if you encounter budgeting difficulties.

Unless you are utterly confident, do not try to achieve too much too early in the year. If you do set out to do this and then fail, you will have to struggle each month to catch up. Nor should you devise a budget where everything is achieved at the end of the year. This can test your fortunes too much. It is not prudent business.

There is good logic to the setting of *target* budgets which overachieve on the annual requirement. These build a safety factor into the system.

People strive harder to reach the *target* goals. You could have a bonus system geared to achieving the targets.

Split the total budget into a series of sector budgets which add up to the total requirement. Each sector will be the responsibility of a separate management team. This will achieve delegation, encourage teamwork, and encourage more people to take budget responsibility — they will fight harder to achieve *their* budget than they will to achieve *your* budget.

There is a further logic in splitting any budget into sectors. No budget will ever be truly perfect, although you can manipulate them to appear so. You will always perform above or below the requirement. A series of budgets, added to give a single total, is likely to be more consistent than a single global shot.

Throughout history, good leaders have kept a reserve for times of need. Be a good leader!

Clearly, budgets are:

- business tools to monitor the business
- usually for a twelve-month accounting period and split into monthly comparisons of cost and value against the budget expectations
- tools which enable the business to identify areas of poor performance. Corrections can then be made which are in the best interests of the business.

What is likely to happen to our budget and what corrective actions are likely?

Costs are greater than expected. This should not occur in a head office department or a consultancy where monthly costs are largely fixed and repetitive. It can occur on a construction site where production costs have been underestimated. This may be due to our own incorrect tender assumptions, ground conditions may be worse than expected, or we may be suffering disruption on site. An awareness of the extent of the problem enables appropriate actions to be planned at an early date. Such action could be a claim or changes to site practices.

Costs are less than expected. This is unlikely to occur in our fixed cost department. It could occur on site, usually because progress is less rapid than expected. We are doing less work in any period and so are likely to use fewer resources — materials are an obvious area. We may need to increase the resources on site. Examination of the construction programme will probably show slippage. Is this our fault?

Cash inflow is less than expected. Address this as necessary. You may have to chase cash owing to the business and control the cash flowing out more carefully.

Cash outflow is greater than expected. Take great care to find the cause and deal with it.

This chapter should have given you a grasp of the budget system. Size and type of value and spend are the only differences in budgets for nations, large and small companies, or your own household. The principles are the same and they are simple.

As with preparing cost estimates, split down large units into smaller ones which *you* can understand and then work on successfully.

Clearly, if you are going to monitor business performance, knowledge of the budget system is necessary.

To conclude what has hopefully been a useful learning experience for the reader, this book outlines two key skills for engineers and managers of all kinds. These are the abilities to:

- Price all productive effort from first principles using relevant costs.
- Understand the principles of how budgets are derived and budgetary control is implemented.

Whilst such abilities may not in themselves transform the reader into a top level manager, the opposite is very likely. Without such abilities you are unlikely to become a top level manager in a successful business.

Index